Cancer Explained

Jon Adams

CONTENTS

INTRODUCTION

"Cancer Explained" unravels the complexity of a subject that, in one way or another, touches the lives of nearly every person on the planet. From its cellular underpinnings to the far-reaching impact it has on individuals and societies, cancer remains one of the most formidable challenges of our time. This book sets out to demystify the disease from A to Z, breaking down the intricate layers of cancer's biology and the nuances of its treatment into comprehensible parts.

Diving into the core of cancer's pathology, the pages of this book illuminate its origins, mechanisms, and progression using simple, clear language. Our journey will traverse the causes and types of cancer, decoding medical terminology and exploring both traditional and cutting-edge treatments. In the chapters ahead, readers will find vivid examples that bring scientific concepts to life, relatable analogies that turn abstract ideas into relatable knowledge, and a narrative rich with insight.

"Cancer Explained" is not just an academic overview; it's an accessible guide crafted with the lay reader in mind. Whether you're a student, a caregiver, someone living with cancer, or a curious mind seeking clarity about this pervasive disease, this book is a beacon of understanding. It's a testament to human resilience, a resource for empowerment, and a wellspring of knowledge thoughtfully designed to engage, educate, and inspire.

Prepare to embark on a journey of enlightenment, one where the intricacies of cancer are no longer shrouded in obscurity but laid bare for all to understand. This is cancer, explained—where every question leads to knowledge, and every fact dispels fear.

CHAPTER 1: THE CONCEPT OF CANCER

Cancer is not a single foe but a collective name for a vast group of diseases that all stem from uncontrolled cell growth. Cells, the fundamental units of life, typically follow a strict code of growth, division, and death. When this code is broken or ignored, a cell may start to divide uncontrollably. The mass of cells that can result from this continuous growth is what we often call a tumor, which can be benign or malignant. Cancer specifically refers to the case of malignant tumors, where cells not only grow without order but also have the potential to invade other tissues, spreading to different body parts through a process known as metastasis.

Understanding cancer begins with grasping these basic processes: how cellular mechanisms fail, leading to disease. It's about observing the body's own systems getting disrupted, causing harm where they once created harmony. Knowing this helps contextualize why treatment options are as diverse as the disease manifestations themselves and underscores the value of research and advancements in the medical field. This understanding brings a sense of clarity and control, highlighting the importance of early detection and individualized approaches in managing health, showcasing the connection between scientific knowledge and personal well-being.

Cancer represents an array of diseases characterized chiefly by cells that behave incorrectly. Unlike normal cells, which grow, divide, and neatly follow their life cycle, cancerous cells break these rules. They replicate rapidly, refuse to stop growing when they should, and fail to self-destruct at the end of their life span—behaviors that healthily functioning cells would typically avoid. Picture a city where traffic lights no longer function as intended, causing chaos on the roads; similarly, cancerous cells create disorder by disregarding the body's biological traffic signals.

At the heart of cancer lies a series of mutations—changes to the DNA, which can be thought of as the cell's instruction manual. These mutations can be random or influenced by external factors like smoking or radiation. As mutated cells replicate, they pass on these errors, accumulating more changes and gaining new, harmful capabilities, such as the ability to invade neighboring tissues or travel to distant sites within the body—a hallmark of cancer's destructive potential.

These abnormal cells form masses called tumors, but not all tumors equate to cancer. Benign tumors, while sometimes problematic, do not spread and are often less concerning. It is the malignant variety, with cells capable of invading other tissues—that earn the cancer label and necessitate more aggressive intervention. Their ability to spread, known as metastasis, is akin to dandelions dispersing their seeds in the wind to take root far from their original location.

One key to combating cancer is to understand this disharmony in cellular behavior. Treatments often focus on rectifying these abnormal actions, akin to reprogramming the errant signals in our city traffic analogy. By recognizing the aberrant behaviors of cancerous cells, we grasp the urgency of research and advancement in treatments, and acknowledge the critical role that early detection plays in managing the disease.

Let's take a closer look at DNA mutations and their role in cancer. DNA can be likened to an instruction manual for cells, directing their growth, function, and death. Within this manual are specific genes called oncogenes and tumor suppressor genes, which are crucial in regulating cell behavior. Oncogenes, when functioning normally, promote cell growth; however, when mutated, they can become overactive, like a gas pedal stuck to the floor, leading to unchecked cell proliferation.

On the other side, tumor suppressor genes act as the brake pedals, halting cell division if something goes awry. Mutations here can disable these brakes, allowing cells to grow out of control. This disruption of cellular equilibrium is often the first step towards the formation of cancer.

As mutated cells divide, they pass on their errors, leading to further genetic instability. Each additional mutation provides the cells with advantageous traits for survival, such as the ability to sidestep programmed cell death or to create their own supply of growth signals. This is where cell signaling pathways come into play. These pathways, a series of chemical reactions within the cell, usually communicate messages such as when to divide or when to die. In cancer, these messages get distorted.

Eventually, as the mutations accumulate, some cells may gain the ability to break away from their original location, invade neighboring tissues, and even travel through the bloodstream or lymphatic system to colonize distant organs – a deadly hallmark of cancer known as metastasis.

Understanding these steps not just in terms of what happens, but how and why, is pivotal. It sheds light on targets for treatment and underscores the critical challenges in cancer therapy. This understanding highlights the intricate complexity of cancer but also brings us closer to the tools needed to combat it. Every detail matters in this intricate story from the molecular-level misconduct that compromises our cellular citizens to the ultimate quest for therapies that can restore order within the body's communities.

Imagine a bustling, efficient city where every citizen and municipal system works in harmony – this is akin to the body's ecosystem when cells are healthy and playing their roles correctly. Now picture a group of residents who, having forgotten the rules of the community, start building wherever they please, disregarding zoning laws and local guidelines. These rogue builders represent mutated cells. They begin constructing structures at an alarming rate, disrupting neighborhoods and the city's infrastructure - similar to how cancerous cells multiply uncontrollably and infringe on bodily tissues.

These rogue elements can be seen as conspirators stirring chaos, causing trouble in the once orderly city by hijacking materials, rerouting traffic, and even enticing other citizens to join their renegade operation. Just as the city would start to fail under the pressure of these unruly activities, the body's ecosystem falters under the strain of mutated cells as they form tumors and potentially spread, encroaching on vital organs much like the unruliness might spread to disrupt neighboring towns.

The significance of this analogy lies in the notion that an understanding and awareness of the residents' responsibilities are akin to the role of cells in the body. Just as peacekeepers or city planners would strive to detect and rectify the initial signs of disorder, medical professionals aim to identify and treat cancer early, restoring harmony and functionality to the body's dynamic and interconnected community.

Here is the comprehensive breakdown of the cellular mechanisms that can lead to cancer:

- Types of DNA damage and cause (e.g., UV radiation, chemicals)

- **Point Mutations:** These occur when a single DNA base pair is altered. It's like a spelling mistake in a word that can change the entire meaning of a sentence.

- **Insertions and Deletions:** Additional base pairs are inserted, or existing ones are deleted, which can scramble a gene's instructions like a misaligned cog in machinery.

- **Effect on DNA Sequence:** Such changes can create faulty proteins or disrupt how genes are read by the cell, akin to corrupt files in a computer system.

- The role of DNA repair mechanisms and failure leading to mutations

- **Cellular Repair Systems:** Cells have mechanisms such as nucleotide excision repair and mismatch repair that act like a quality control team, fixing errors before they become problematic.

- **Overwhelmed or Defective Repairs:** When damage outpaces repair, or the repair systems themselves are faulty due to mutations, errors accumulate like unchecked corrosion in a structure.

- Activation of oncogenes and deactivation of tumor suppressor genes

- **Triggers for Oncogenes:** Factors like viruses or carcinogens can permanently switch on genes that promote growth, much like a car's accelerator gets stuck, leading to unstoppable speed.

- **Tumor Suppressor Genes 'Off':** Damage or mutation can switch off genes meant to slow down proliferation, like cutting the brakes of a vehicle, leading to loss of control over cell growth.

- Detailed process of how normal cell functions are altered by mutations

- **Pathway to Uncontrolled Division:** Mutated cells neglect the signals that dictate normal division and not only multiply at an increased rate but also avoid the natural process of cell death, continuing to accumulate.

- **Evasion of Death and Growth Signals:** They may produce false signals to protect themselves against cell death or ignore the body's stop signs that regulate growth, resulting in a dangerous free-for-all.

- The transition from localized to invasive cancer

- **Metastasis Steps:** Cells may undergo changes in their structure allowing them to detach and invade into neighboring tissues, travel through the blood or lymphatic systems, and establish colonies in new environments, far from their original site.

- **Cellular and Tissue Changes:** Just as an aggressive plant species might spread through a field, these cells alter their surroundings, breaking away and seeding themselves in different parts of the body.

Each step in this process adds a piece to the complex puzzle of cancer formation. Understanding these components provides insights into not just how cancer arises, but also how it might be detected and treated. Though the content is intricate, it's imperative for us to grasp these details to appreciate the full scope of cancer as more than a mere 'disease' but as a series of cellular events with profound implications for human health.

In the process of genetic mutation that leads to cancer, the DNA within a cell, which contains the instructions for building and maintaining the body, experiences changes in its sequence. Normal cells have a set of checks and balances that control growth and division, ensuring they function correctly. However, when errors occur in the DNA sequence, it can disrupt this balance.

The process begins with a change in one of the many genes that regulate cell growth. Such changes can occur randomly or due to external influences like tobacco smoke or UV radiation. When a gene that promotes cell growth (known as an oncogene) is altered, it might become overactive, causing the cell to multiply more than it should. Conversely, when a gene responsible for suppressing growth (a tumor suppressor gene) is altered, it might lose its ability to control cell division.

Cells have mechanisms to repair DNA errors, but these can fail, especially if the damage is extensive or the repair systems are themselves mutated. This failure allows further mistakes to accumulate every time the cell divides. As mutations build up, the cell gradually diverges from its normal state and begins to divide uncontrollably, leading to the growth of tumors.

Some cancer cells may gain the ability to move and invade other parts of the body, spreading the disease. This situation is particularly grave because it disrupts the function of multiple organs and complicates treatment. Understanding this mutation process is crucial as it underpins the strategies for cancer prevention, diagnosis, and therapy. Each mutation is a potential target for intervention—blocking the overactive oncogenes or restoring the function of tumor suppressor genes could help in treating cancer.

Cancer development is a complex multi-step process that begins at the most basic unit of human life: the cell. Let's unpack this process step by step to understand how a healthy cell turns into a cancer cell.

First, our DNA is subject to damage every day, which can be induced by various factors including UV radiation, smoking, or even errors during cell division. Our cells, however, are equipped with repair systems, akin to proofreaders that correct errors in text. These systems, which include mechanisms like nucleotide excision repair and base excision repair, scrutinize the DNA and fix the damage.

However, if the damage is extensive or the repair systems are faulty, errors accumulate in the DNA. This is where mutations come into play. Mutations are permanent alterations in the DNA sequence. Think of them as typos in a critical essay - some might go unnoticed, while others can change the entire meaning. There are several types of mutations that can lead to cancer:

- **Missense Mutations**: These are like typo errors that result in the wrong word being used in a sentence. In DNA, this leads to a change in one amino acid, the building block of proteins, which can affect the protein's function.

- **Nonsense Mutations**: Nonsense mutations are the equivalent of a full stop placed in the middle of a sentence. They create a signal that prematurely stops protein production, leading to truncated, nonfunctional proteins.

- **Frameshift Alterations**: These mutations are like adding or removing a space in a sentence, shifting the grouping of letters and rendering the text nonsensical. In DNA, these mutations shift the "reading frame," often resulting in a completely dysfunctional protein.

Once mutated, genes that regulate the cell cycle can become

11

dysfunctional. Oncogenes, which could be compared to cell growth promoters, can be locked in an 'on' position, leading to constant growth signals. Tumor suppressor genes, which normally put the brakes on growth, can lose their ability to perform this vital function.

As more mutations accumulate, a cell may start to grow uncontrollably and become what we call a tumor. Initially, this growth is localized. If these cells acquire the ability to invade adjacent tissues, we begin to see the transition from a localized tumor to invasive cancer.

In the bloodstream, cancer cells can travel to distant tissues, a process known as metastasis. It's like them finding new places to "live" within the body, which makes treatment much more challenging.

Modern treatments aim to target these specific breakdowns in cellular regulation and function. For example, targeted therapies are designed to switch off oncogenes or reactivate tumor suppressor genes. Other approaches, like immunotherapy, empower the body's own immune system to recognize and destroy cancer cells.

By understanding this step-by-step process, we can begin to appreciate the complexity of cancer and the importance of advancements in both our understanding and treatment of this disease. Each mutation and cellular change represents a potential target for intervention, leading to better outcomes for patients.

This narrative presents a focused exploration of cancer's cellular mechanics for those who yearn to understand without wading through a sea of scientific jargon. The goal is to take something intricate and make it not just comprehensible but also applicable to real-world contexts, guiding through the information methodically and systematically.

Picture a serene forest where each tree grows at a measured pace, contributing to the balanced ecosystem. In the midst of this forest, imagine a single tree begins to grow at an unprecedented rate. It draws more water and nutrients than it needs, depriving the surrounding flora. This tree represents a tumor, with its growth mirroring the unchecked proliferation of cancer cells.

As this tree expands, it sends out roots far and wide, unsettling the soil and disrupting the harmony of the forest floor. These invading roots mirror the way cancer cells infiltrate and wreak havoc upon neighboring tissues in the body. Furthermore, imagine seeds from the overgrown tree carried off by the wind to take root in distant parts of the forest, akin to cancer cells traveling through blood or lymph to establish new tumors in other organs.

This escalating growth and invasion demonstrate not only how tumor cells operate but also underscore the importance of maintaining the forest's—or body's—natural balance. Just as foresters might seek to curtail the spread of our hypothetical tree, medical professionals work to halt the progression of cancer, preserving the health of the entire organism. This analogy simplifies the complexity of tumor growth and spread, grounding it in a relatable narrative that captures both the disruptive nature of the disease and the crucial efforts to control it.

Here is the detailed breakdown on the alterations in the cell cycle that lead to cancer growth:

- Disruptions in the Cell Cycle Leading to Proliferation
- Normal cells follow a series of steps in the cell cycle to grow and divide orderly.
- Cancerous cells bypass these controls, continuing to divide without the usual signals.
- This disruption can be driven by mutations that damage DNA, causing an accelerated cycle and increased cell numbers.

- The Role of Proto-oncogenes and Oncogenes
- Proto-oncogenes are genes that, when functioning normally, help cells grow.
- When mutated, these proto-oncogenes become oncogenes, which can cause uncontrolled cell growth.
- Oncogenes can be likened to a gas pedal stuck down, propelling the vehicle (the cell) forward without restraint.

- Function of Tumor Suppressor Genes like p53
- Tumor suppressor genes act as the brakes for cell growth and division.
- The p53 gene monitors the cell for DNA damage and can trigger repair

or cell death if necessary.

- When tumor suppressor genes are inactivated, cells can grow unchecked, like a car without brakes.

- Angiogenesis: Blood Supply for Tumors

- Tumors need a supply of nutrients and oxygen to grow, which they achieve through angiogenesis.

- Angiogenesis creates new blood vessels to nourish the tumor, similar to laying down new pipes to a building to ensure its functions.

- Mechanisms of Cellular Immortality

- Telomeres protect the ends of chromosomes and normally shorten with each cell division, eventually leading to cell death.

- Cancer cells often activate an enzyme called telomerase, which maintains telomere length, allowing the cell to divide indefinitely.

- Cellular Adaptations for Metastasis

- Cancerous cells adapt to detach from the original tumor, enter the bloodstream or lymphatic system.

- They must survive the journey through the bloodstream, which is hostile to most cells.

- Once they reach a new location, they invade the tissue and begin to grow a new tumor.

By understanding these points, readers can appreciate the complex series of events that transform normal cells into cancerous ones. Each component, from mutation to metastasis, plays a pivotal role in the development and progression of cancer, offering potential targets for therapeutic intervention. It is through this detailed comprehension that the importance of research and innovation in cancer treatment becomes clear.

Cancer, a disease that does not discriminate, has touched the lives of people from all walks of life, including those in the public eye. Take, for example, the beloved children's author Roald Dahl, who battled with leukemia for a significant period. Through his ordeal, he maintained a sense of wonder and creativity that continued to imbue his work, proving that even amid personal struggles, the human spirit can thrive. His story resonates with resilience and the capacity to find joy and purpose in the face of adversity.

Similarly, acclaimed Apple Inc. co-founder Steve Jobs' bout with a rare form of pancreatic cancer is well-documented and has raised broad awareness about the need for research and innovation in cancer treatments. Jobs' journey with cancer was one of determination and focus, and his legacy includes not just technological advancements but also a candid narrative of his fight against this pervasive disease.

In both cases, the impact of cancer extends beyond the individual to touch the lives of fans and followers, illustrating the universal reach and deep emotional resonance of this condition. These stories reflect not just personal battles but also the collective call to action to better understand and ultimately conquer cancer. They provide tangible examples of courage and serve as a poignant reminder of the threads of vulnerability and strength that connect all humans.

In navigating the complexities of cancer, we have unpacked its many facets: the cellular anomalies that ignite its onset, the genetic mutations that drive its progression, and the physiological impacts that underscore its severity. Each component examined has shed light on the inner workings of a disease often shrouded in uncertainty, painting a clearer picture of its nature.

Understanding these concepts is more than academic; it equips you with the knowledge to engage with cancer on a more informed level. Comprehending the why and how of cancer cell behavior demystifies the disease, transforming fear into awareness. It becomes a source of empowerment, offering the reader a foundation upon which to build an informed perspective of the condition, whether it's for personal enlightenment, supporting a loved one, or contributing to the broader dialogue on health.

This chapter concludes not at the end of a journey but at the beginning of embracing a deeper awareness. By recognizing the intricacies of cancer's development, one can better appreciate the strides made in treatment and the promise of future discoveries. It's this understanding that fosters an informed and hopeful outlook and fortifies the collective resolve to continue pressing forward in the quest for cures.

CHAPTER 2: ANATOMY OF A CELL

Every living organism is composed of building blocks known as cells; they are as complex as they are essential, forming the image of life itself. To understand the cell is to gain insight into the very essence of human health and the myriad diseases that can afflict us—including cancer. Cells are miniature powerhouses, each harboring a universe of structures, each with a role critical to the organism's survival. The anatomy of a cell includes components that provide energy, synthesize proteins, and command growth and repair. Each part must function correctly, for even the slightest deviation can lead to significant consequences. As we look at the cell's intricate anatomy, we step into a world where clarity bridges the gap between the known and the mysterious, guiding us through a microcosm that mirrors the complexity of life itself. This understanding is not only academic; it is a lens through which we can view the landscape of life, disease, and the continuum of health that connects us all.

For a second I want to you to imagine entering a grand building with a discerning doorman who decides who can come in and who stays out. Similarly, the cell membrane functions as the vigilant gatekeeper of the cell, a dynamic barrier that dictates what substances are allowed entry and which are denied. Like the skilled doorman, the cell membrane oversees the flow of materials, selectively permitting the ingress of vital nutrients and the egress of waste products. This selective permeability is key not only to sustaining the life of the cell but also to maintaining the entire organism's equilibrium.

By controlling what passes through, the membrane preserves the cell's internal environment, ensuring that the conditions inside remain optimal for the cell's intricate operations to proceed without hitch. In the same way that a secure, well-managed building creates an exclusive, stable environment, the integrity of the cell membrane maintains the balance critical for the cell's function and, consequently, our health. Understanding the part the cell membrane plays sheds light on the fundamental processes that underpin our very existence and offers a window into the delicate balance of life at the cellular level.

Let's look at the breakdown on the cell membrane's structural features and physiology:

- The Composition and Fluid Nature of the Phospholipid Bilayer
- Composed of two layers of phospholipids, molecules with a hydrophilic (water-attracting) head and two hydrophobic (water-repelling) tails.
- The bilayer is fluid, allowing lateral movement of proteins and lipids, much like icebergs drifting in the sea.
- This fluidity is critical for the functioning of certain proteins and for the fusion and fission of membranes.

- Types of Membrane Proteins and Their Roles
- Integral proteins span the membrane and can act as channels or pumps to move substances across the barrier.
- Peripheral proteins are attached to the surface of the membrane and play a role in signal transduction and cell communication.
Amplifiers of cell signals, relaying messages from the body to the inside of the cell.

- Mechanisms of Selective Permeability
- Passive transport, such as diffusion, allows substances to move across the membrane without the use of cellular energy based on concentration gradients.
- Active transport requires energy, typically in the form of ATP, to move substances against their concentration gradient.
- These processes ensure that essential nutrients enter the cell while waste products are expelled, maintaining homeostasis.

- The Role of Cholesterol in Membrane Integrity and Fluidity
- Cholesterol molecules intersperse among phospholipids, modulating the fluidity of the membrane.
- They stabilize the membrane in varying temperatures by preventing the fatty acid tails from coming too close or becoming too rigid.

- The Concept of Membrane Carbohydrates and Their Function
- Membrane carbohydrates are often attached to proteins (glycoproteins) or lipids (glycolipids).
- They are key in cell recognition and signaling, serving as identification

tags that enable cells to communicate and coordinate with each other.

Understanding the cell membrane's makeup and functions clarifies its complex role as more than a mere barrier—it is an active and dynamic interface critical to the life of the cell. Its components work synergistically to create an environment where exchange and communication happen effectively and efficiently. Through this comprehension, we admire the elegance and precision of this living boundary that is fundamental to both the survival of individual cells and the health of the entire organism.

The nucleus serves as a command center for the cell, safeguarding the integrity of genetic information and coordinating key functions. Encircled by a double membrane, the nucleus houses DNA, the instruction manual for all cellular operations, akin to a central database containing a collection of blueprints. Within this database, DNA is organized into chromosomes when the cell prepares to divide, ensuring each daughter cell inherits complete and accurate genetic instructions.

Functioning in harmony with the rest's cellular machinery, the nucleus not only protects DNA but also mediates the transcription process, whereby specific segments of DNA are copied into RNA. This RNA then travels outside the nucleus, guiding the synthesis of proteins in cytoplasmic structures known as ribosomes. Through this process, called gene expression, the information stored in the nucleus translates into tangible action, directing everything from cell metabolism to growth and repair.

The nucleus also plays a pivotal role in the life cycle of a cell, overseeing cellular division and reproduction. It initiates and controls the delicate phases of the cell cycle, a series of steps ensuring that the genetic material is duplicated correctly and distributed evenly during cell division.

Each aspect of the nucleus's function is crucial for maintaining life at the cellular level. Understanding the nucleus's role clarifies the larger picture of cellular life, offering a sharper image of the balance and complexity intertwined within each living organism. This knowledge not only informs but also equips us with a deeper connection to the microscopic world that thrives within us all.

Let's take a closer look at the nucleus, the central hub of the cell that manages genetic information and orchestrates key functions. The nucleus is encased in the nuclear envelope, a structure comprising inner and outer membranes. This envelope acts as a security checkpoint between the nucleus and the rest of the cell, complete with nuclear pores that regulate the flow of molecules. These pores are the channels through which vital substances like proteins, RNA, and signaling molecules pass, ensuring communication and transport are both controlled and efficient.

At the nucleus's heart lies the nucleolus, a densely packed region where the assembly of ribosomes begins. The nucleolus is a focal point for producing ribosomal RNA (rRNA), a fundamental component of these cellular machines that construct proteins. Within the wider nuclear space, DNA is meticulously organized into chromatin, which condenses to form chromosomes when a cell prepares for division. Two forms of chromatin coexist: euchromatin, loosely packed and actively involved in gene expression, and heterochromatin, more condensed and typically less active.

Chromosomes occupy specific areas known as chromosome territories, which influence how genes are accessed and regulated. This spatial organization within the nucleus matters because it affects which genes are active and which remain silent, this playing a crucial role in a cell's function and identity.

Moreover, the nucleus is central to the cell cycle's orchestration, operating checkpoints that assess whether DNA has been replicated without errors before cell division proceeds. It acts as a quality control manager, ensuring the genetic information passed to successor cells is intact and free from damage.

By understanding these intricate nuclear components and their staggering coordination, we appreciate the genius behind cellular mechanics that occur on a microscale but bring about significant effects on life as we know it. This transparent glimpse into the nucleus's inner workings exemplifies how monumental complexities can be mastered through thoughtful and clear examination, mirroring the elegance inherent in all biological systems.

Think of the mitochondria as the miniature power stations of the cell. Just as a power plant burns fuel to produce energy for a city's needs, keeping the street lights bright and machinery humming, mitochondria convert nutrients into ATP, which is the primary energy currency of the cell. This process is akin to generating electricity; through a series of biochemical reactions known as the Krebs cycle followed by the Electron Transport Chain, mitochondria take in 'fuel' in the form of sugars and fats and release usable energy.

Within each mitochondrion, the inner membrane plays host to these energy-producing reactions, much like the inner workings of a power plant where combustion occurs. From here, ATP is distributed throughout the cell, powering everything from muscle contractions for the beating of your heart to the firing of neurons that allow you to appreciate a beautiful sunset.

This is why the role of mitochondria is so pivotal; without them, cells wouldn't have the energy required to perform life-sustaining tasks. The efficiency and effectiveness with which mitochondria produce and dispatch energy underscore their importance, not just for individual cells but for the health and vitality of the entire organism. Understanding their function provides insight into how our bodies operate at the most fundamental level— the conversion of the food we eat into the energy that fuels our lives.

Here is the breakdown on the detailed mitochondrial processes involved in ATP production:

- Inputs and Outputs of the Krebs Cycle
   - Inputs include acetyl-CoA derived from carbohydrates, fats, and proteins.
   - Outputs are ATP, reduced coenzymes NADH and FADH2, and waste product carbon dioxide.
   - Key intermediates like citrate, alpha-ketoglutarate, and succinate are involved in a cyclical sequence of enzyme-driven reactions.

- Electron Transport Chain
   - Consists of a series of protein complexes (I-IV) embedded in the inner mitochondrial membrane.
   - NADH and FADH2 donate electrons to the chain, which are passed along via redox reactions.
   - The movement of electrons creates a proton gradient across the membrane, with protons accumulating in the intermembrane space.

- Oxidative Phosphorylation

- ATP synthesis occurs when protons flow back into the mitochondrial matrix through ATP synthase due to the established gradient.

- ATP synthase acts as a molecular turbine, using the energy from the protons' movement to convert ADP and inorganic phosphate into ATP.

- Mitochondrial DNA Regulation

- Mitochondria possess their own circular DNA, which is independent of the cell's nuclear DNA.

- Mitochondrial DNA contains genes essential for the organelle's function, primarily coding for enzymes required in oxidative phosphorylation.

- The maternal inheritance of mitochondrial DNA has unique implications for genetics and diseases.

- Mitochondrial Dynamics

- Mitochondria are dynamic; they continually undergo fission (division) and fusion (joining), enabling them to change shape and quantity in response to cellular energy demands.

- These processes are crucial in maintaining mitochondrial function and quality control, affecting overall cellular health.

By deeply understanding each of these components and their functions, one gains a clear view of just how mitochondria act as the powerhouse of the cell—efficiently orchestrating a complex series of events to produce life's essential energy molecule, ATP. It is this understanding that not only fascinates but emphasizes the centrality of mitochondria in the life of every cell, and ultimately, in the life-sustaining processes of our bodies.

Ribosomes are the cellular machines responsible for protein synthesis, a process essential for countless activities within the body, from muscle contraction to immune response. One can think of ribosomes as small but highly efficient assembly lines, where the instructions from the DNA are translated into proteins—the workhorses of the cell.

These instructions, transcribed into messenger RNA (mRNA) in the nucleus, are delivered to the ribosomes stationed either freely in the cytoplasm or attached to the endoplasmic reticulum. Ribosomes read the sequence of the mRNA and, with the assistance of transfer RNA (tRNA), which brings specific amino acids, they chain these building blocks into polypeptide chains following the order dictated by the mRNA—a process known as translation. The sequence in which amino acids are assembled determines the protein's unique structure and function.

This production line operates with precision, starting when the ribosome assembles around the initiation point on the mRNA and continuing until it reaches a stop codon, signaling that the complete protein has been synthesized and can be released to perform its designated function.

As we look at the role of ribosomes, we can reveal a fundamental aspect of cellular biology, illustrating how the decoding of genetic information results in the formation of proteins, essential players in the drama of life. Each step in this translation process is meticulously choreographed and crucial for sustaining the body's functions, highlighting the extraordinary efficiency encapsulated in these microscopic entities. Understanding ribosomes and protein synthesis opens a window to a deeper understanding of foundational biological processes with wide-reaching implications in health, medicine, and even biotechnology.

Protein synthesis, orchestrated by ribosomes, is a finely tuned process that can be broken down into several distinct phases. The story begins in the initiation phase, where the small subunit of a ribosome attaches to a messenger RNA (mRNA) molecule. It scans along the mRNA to find a start codon, which is typically AUG. This codon signals the start of a protein-coding region. The large ribosomal subunit then joins to form a functional ribosome, ready for the task ahead.

Next unfolds the elongation phase. During this stage, transfer RNA (tRNA) molecules, which can be thought of as adaptor molecules, bring amino acids to the ribosome. Each tRNA has an anticodon that pairs with a corresponding mRNA codon, ensuring the correct sequence of amino acids. Peptide bonds form between amino acids, linking them into a growing polypeptide chain. This intricate dance, where the ribosome moves from one codon to the next, adding amino acids to the polypeptide chain, repeats continuously.

Central to this process is ribosomal RNA (rRNA), a type of RNA that, along with proteins, forms the ribosomal subunits. Ribosomal RNA catalyzes the formation of peptide bonds, a reaction crucial to linking amino acids.

The performance reaches its finale in the termination phase. When the ribosome encounters a stop codon on the mRNA—UAA, UAG, or UGA—this signals that the protein synthesis is complete. At this cue, the ribosome halts synthesis, dissociates into its subunits, and releases the newly synthesized polypeptide.

After release, the polypeptide chain undergoes folding into a specific shape, which is imperative for the protein's function. In some cases, additional modifications such as the addition of carbohydrates or lipids are necessary for the protein to become fully functional.

This entire process is tightly regulated within the cell. Various factors influence the efficiency and accuracy of protein synthesis, ensuring that proteins are produced as needed, without errors that could lead to dysfunction or disease. The complexities involved in the translation process highlight the ribosome's critical role in maintaining the health of cells and the organism as a whole. Through a deeper understanding of this process, one can truly appreciate the intricacy and efficiency of life at the molecular level.

Picture the endoplasmic reticulum (ER) as the bustling assembly line of the cell, not unlike a vast and intricate factory floor where a myriad of products are methodically put together and prepared for shipping. This network of membranous channels takes on the critical task of synthesizing proteins, much like workers on an assembly line meticulously adding parts to create a final product. Ribosomes, acting as specialized workstations, are attached to the rough ER and are where the building blocks of proteins are sequentially assembled, following the instructions encoded in mRNA.

Once the proteins are assembled, they are folded into their active forms with the help of chaperone molecules, akin to quality control ensuring that each item is up to standard. If proteins need to be delivered externally or to specific compartments within the cell, the smooth ER, devoid of ribosomes, takes over. It packages these proteins into vesicles—little bubble-like structures—that are then dispatched to their destination, similar to a warehouse packaging department preparing goods for transport.

The elegance of the ER lies in its dual roles and its adaptability; it can switch functions between protein synthesis and shipping. This compartmentalized approach within the cell exemplifies the efficiency of cellular processes and highlights the importance of the ER in maintaining the functional integrity of the cell. Understanding the ER is not just an intellectual exercise—it brings into proximate view the dynamic and harmonious nature with which cells preserve life on a microscopic scale, which in turn, reflects the marvel of biological systems and their evolution.

Here is the detailed list of the endoplasmic reticulum's role in cellular processes:

- **Rough ER and Protein Synthesis**
  - Covered with ribosomes, rough ER is responsible for synthesizing proteins destined for the cell membrane, outside of the cell, or for specific cellular compartments.
  - As proteins are synthesized by ribosomes, they enter the rough ER lumen where they begin to fold into their functional structures.

- **Co-Translational Translocation**
  - During protein synthesis, some proteins are marked by a signal peptide, directing them to the rough ER.
  - The signal recognition particle (SRP) pauses translation and guides the ribosome-protein complex to the ER membrane where the protein enters.
  - Inside, signal peptidase removes the signal peptide, and protein synthesis continues with the nascent chain entering the ER.

- **Smooth ER and Lipid Metabolism**
  - Lacking ribosomes, the smooth ER is involved in lipid and steroid hormone synthesis.
  - It also plays a role in detoxifying metabolic byproducts, drugs, and poisons.

- **Molecular Chaperones and Protein Folding**
  - Inside the ER lumen, molecular chaperones assist in protein folding and assembly.
  - Proper folding is critical for protein function and is meticulously monitored.

- **Vesicular Transport**
- Vesicles are small sacs that transport proteins from the ER to other parts of the cell.
- COPII vesicles move materials from the ER to the Golgi apparatus, while COPI vesicles shuttle them back.

- **ER Quality Control and the Unfolded Protein Response (UPR)**
- The ER has a quality control system that ensures only properly folded proteins move to the next station.
- When misfolded proteins accumulate, the UPR is activated to restore normal function by slowing protein synthesis and increasing chaperone production.
- Persistently misfolded proteins are tagged for degradation by the cell.

Through a straightforward explanation of these complex processes, one can appreciate how the ER functions like an efficient factory within the cell, crafting and shipping essential molecular machinery. Understanding these fundamental operations is critical for recognizing the remarkable efficiency and coordination of cellular life, showcasing the integral role the ER plays in maintaining health at the microscopic level.

If the endoplasmic reticulum is the cell's factory line, then the Golgi apparatus is its bustling post office. Here, in this central dispatch center, proteins that have been previously assembled and folded in the ER are received and processed. The Golgi apparatus meticulously sorts these items, much like a postal worker would categorize parcels, determining their final destinations.

But it doesn't just sort; the Golgi apparatus also puts the final touches on proteins and lipids, adding molecular 'address labels' in the form of carbohydrate groups that ensure delivery to the right location, akin to affixing postage stamps on mail. Some of these proteins and lipids might be destined for the outside world – secreted out of the cell, while others could become part of the cell membrane or delivered to various compartments within the cell.

Just like a loading dock of a shipping center, the Golgi apparatus packages these products into vesicles, small bubble-like structures that are capable of traveling great intracellular distances. Each vesicle is carefully packaged and labeled for its specific journey, ready to be transported along the cytoskeleton's network – the cell's own highways.

Understanding the Golgi apparatus's roles aligns us with a deeper respect for the cell's internal logistics, showcasing a level of biological organization and precision that is crucial for the survival of the cell and, by extension, the organism. This microscopic freight system emphasizes how even the smallest components within us operate with a finesse and sophistication that is nothing short of remarkable.

Here is a quick breakdown of the complex yet fascinating activities within the Golgi apparatus:

- **Structural Organization of the Golgi Apparatus**
- Divided into regions known as the cis-Golgi network, cis, medial, and trans cisternae, and the trans-Golgi network, each with unique enzymatic makeup for different stages of processing.
  - The cis face receives proteins from the endoplasmic reticulum.
  - The medial cisternae are where most of the protein modification takes place.
  - The trans cisternae dispatch processed proteins to their destinations.
  - The trans-Golgi network acts as the final sorting and dispatch area.

- **Types of Enzymatic Modifications**
- Glycosylation, where sugars are added to proteins in a stepwise fashion, critical for proper folding and function.
- Phosphorylation, which can tag proteins for localization in the cell, such as lysosomes.

- **Vesicle Formation Mechanisms**
- Coats, such as COPI and COPII, help shape the membranes around vesicles and select components for transport.
  - The COPII coat is involved in moving materials from the ER to the Golgi.
  - The COPI coat often works in retrograde transport, moving substances from the Golgi back toward the ER or within Golgi compartments.
  - Cargo receptors recognize and trap specific proteins meant for

26

transport to different cellular destinations.

- Vesicle Targeting Process
- Vesicles are directed to specific locations based on their coating and cargo receptors.
- SNARE proteins on vesicle and target membranes interact and facilitate the precise fusion of vesicles with their target membranes.

- Regulation of Golgi Function
- The Golgi apparatus can sense and respond to changes in cellular activity, such as the need for increased protein secretion.
- Under stress conditions, it can alter its processes to deal with misfolded proteins or trafficking disruptions.

By diving into the Golgi apparatus's complex inner workings and revealing its efficiencies, we witness a cellular logistics center operating with precision and adaptability. This microscopic postal service is not only integral to the delivery of cellular products but also to the balance and functionality of every living cell it serves. Understanding these details allows one to gain a deeper appreciation for the cell's internal environment, where every component is vital to the seamless continuation of life's processes.

In each cell, lysosomes and peroxisomes act as specialized compartments for waste management and detoxification. Lysosomes are akin to recycling centers; they are filled with enzymes capable of breaking down biological polymers—such as proteins, nucleic acids, and complex sugars—into their monomeric units. This breakdown process allows for the reuse of these materials and ensures that cellular debris does not accumulate to toxic levels.

Peroxisomes have a slightly different function. These organelles contain enzymes that neutralize toxic compounds, particularly hydrogen peroxide, a byproduct of cellular metabolism that can cause significant damage if allowed to accumulate. Peroxisomes convert hydrogen peroxide into water and oxygen, a process comparable to a detox unit neutralizing pollutants to safeguard the environment.

Both lysosomes and peroxisomes are integral to the cell's health and functionality. By sequestering harmful substances and breaking down excess or damaged organelles, they contribute to the cell's longevity and prevent damage that can lead to cellular dysfunction. Just as a city relies on both recycling centers and waste treatment plants to manage its waste and keep its environment clean, cells depend on these microbodies to manage their own internal ecosystem efficiently. Through understanding the functions of lysosomes and peroxisomes, one gains a glimpse into the cell's ability to sustain itself and thrive amidst the complexity of biological processes.

Let's take a deeper look at the sophisticated inner workings of lysosomes and peroxisomes:

- ### Hydrolytic Enzymes in Lysosomes
  - A plethora of enzymes exist within lysosomes, including proteases, lipases, glycosidases, and nucleases.
    - Proteases target proteins, breaking them down into amino acids.
    - Lipases disassemble lipids into fatty acids and glycerol.
    - Glycosidases cleave sugars from glycoproteins and glycolipids.
    - Nucleases degrade nucleic acids (DNA and RNA) into nucleotides.

- ### Autophagy and Lysosomes
  - During autophagy, damaged organelles or unused proteins are enclosed in a membrane to form an autophagosome.
  - This autophagosome then fuses with a lysosome, where the contents are broken down and recycled.

- ### Biochemical Processes in Peroxisomes
  - Peroxisomes carry out beta-oxidation, where very-long-chain fatty acids are shortened for subsequent metabolism.
  - They also break down excess purines into urea, a process essential for purging the cell of toxic build-ups.

- ### Plasmalogens in Peroxisomes
  - Plasmalogens are synthesized in peroxisomes and are vital for the structural integrity and function of cell membranes.
  - The enzymes responsible for their biosynthesis and degradation are specifically located within these organelles.

- ### Reactive Oxygen Species and Antioxidative Mechanisms

- Peroxisomes generate hydrogen peroxide as a byproduct of fatty acid metabolism which is then converted into water and oxygen by catalase.
- The balance between the generation of reactive species and their detoxification is crucial to prevent oxidative stress.

- Implications of Functional Defects
- Dysfunctions in lysosomal enzymes can lead to lysosomal storage disorders, where undigested materials accumulate, impeding normal cell function.
- Peroxisomal disorders, such as Zellweger syndrome, are caused by mutations that affect peroxisome biogenesis, leading to developmental and metabolic abnormalities.

By exploring these details, one can appreciate the nuanced and vital roles of lysosomes and peroxisomes in maintaining cellular integrity. The disruption of these organelles' functions can lead to profound health consequences, making their study not only a fascinating scientific inquiry but also a crucial endeavor with potential therapeutic outcomes. Understanding these cellular components in such depth brings us closer to grasping the delicate and complex nature of human biology.

Envision the cytoskeleton as the complex web of roads, bridges, and frameworks that underpin a bustling city. This intricate network inside the cell provides much the same support and transport pathways as a city's infrastructure does for its inhabitants. The cytoskeleton gives cells their shape and structure, just as a building's framework defines its silhouette against the skyline.

Within this cellular city, microtubules act as highways on which cargo is shuttled back and forth, aligning much like train tracks for the delivery of crucial molecules throughout the cell. Intermediate filaments provide resilience and tensile strength, akin to steel beams in skyscrapers, enabling cells to withstand external pressures and stresses. Meanwhile, actin filaments can be likened to busy streets that allow for the movement of materials and even contribute to the cell's movement and division, similar to traffic flowing through city roads.

Collectively, this cytoskeletal network not only maintains the physical order within the cell's confines but also organizes the flow of traffic, ensuring that all components work in a synchronized fashion—much like a well-planned urban environment. By understanding the cytoskeleton's role in cellular structure, transport, and dynamics, we gain insight into the cell's inner workings that reflect the intricacy and elegance of biological design, reminiscent of our own engineered marvels. This comprehension does more than delineate functions; it fosters an appreciation for the miracle of life that plays out at microscopic scales.

Let's look at the breakdown on the fascinating components of the cytoskeleton and their varied roles within the cell:

- Microtubules
- Composed of alpha and beta tubulin subunits that assemble into a tube-like structure.
- Possess dynamic instability, characterized by alternating growth and shrinkage, essential for cell shape and distribution of organelles.
- Powered by motor proteins like kinesin and dynein, which carry cargo along the microtubules, in a manner similar to trucks on a highway, enabling efficient intracellular transport.

- Intermediate Filaments
- Comprise several types including keratins, which are abundant in epithelial cells, and lamins that form a meshwork beneath the nuclear envelope.
- Provide mechanical strength to cells and help maintain their structural integrity, particularly under stress conditions.

- Actin Filaments
- Feature a distinct polarity with a plus and a minus end, allowing them to rapidly assemble and disassemble.
- Interact with myosin motor proteins, facilitating muscle contraction and playing a crucial role in cell movement and shape changes.

- Cytoskeleton and Cellular Signaling
- Participate in signal transduction processes, serving as scaffoldings for the assembly and operation of multiprotein complexes.
- Change their dynamics in response to various stimuli, influencing cell signaling pathways and affecting cellular behavior.

- ## Cytoskeleton in Cell Division
    - During mitosis, microtubules organize into the mitotic spindle, a structure critical for aligning and separating chromosomes.
    - Ensure that each new daughter cell inherits an identical set of chromosomes, safeguarding genetic information across generations.

These features highlight the cytoskeleton's indispensable role as the cell's all-encompassing framework, influencing not only cell shape and support but also fundamental aspects of cellular transport, signaling, and replication. Understanding the cytoskeleton in these detailed dimensions allows those interested to comprehend how it underpins the cell's operations, affecting its interaction with the environment and contributing to the larger functions of tissues and organs.

In this chapter, we've pieced together the complexities of cellular anatomy, revealing that each component has a distinct yet interdependent function vital to the life of a cell. To start, the cell membrane operates as a selective barrier, regulating the entry and exit of substances in a manner that preserves the cell's internal conditions. Next, we observed the nucleus, the mastermind holding genetic blueprints that dictate the cell's activities and drive protein synthesis through the intermediary, messenger RNA.

The rough endoplasmic reticulum, dotted with ribosomes, emerged as a manufacturing site for proteins, while the smooth variant specializes in lipid production and detoxification. Meanwhile, the Golgi apparatus operates as a cellular post office, processing and tagging molecules for delivery. We also noted the powerhouses of the cell—the mitochondria—where biochemical processes convert energy into a form usable by the cell.

Further, the lysosomes and peroxisomes are like waste disposal units, breaking down macromolecules and neutralizing toxic substances, respectively. Zooming out, we see the cytoskeleton, providing structural support and facilitating movement both within and outside the cell, similar to a city's transport infrastructure.

Collectively, these components sustain vital cell processes, from energy production and biomolecule synthesis to transport, communication, and waste management. Their seamless interaction is crucial not just for

31

individual cells but for the well-being of the entire organism. Understanding these parts in concert showcases the remarkable efficiency and synchronization necessary for life to persist at the microscopic level.

CHAPTER 3: WHEN CELLS REBEL

Chapter 3: When Cells Rebel, looks closely at the complex sequence of events that corrupt the normal lifecycle of our cells, leading them into a state of rebellion we know as cancer. Normally, cells reproduce in a controlled fashion, but as mutations accumulate, this control is undermined, setting the stage for unchecked division and growth. This chapter systematically unveils each step of this mutiny, from the initial genetic alterations to the eventual invasion of tissues, without veiling the facts in medical jargon. Just as a software bug can cause a well-behaved application to behave erratically, so too can genetic mutations prompt cells to deviate from their usual constraints. The ramifications of this cellular insurgence are immense, not only for the individual cell but for the entire organism, leading to a battle for supremacy within its own body. Here, we'll dissect these processes, casting light on the intricate ballet of life, mutation, and survival within our cellular world.

The cell cycle, a vital process for growth and healing, operates under stringent control, akin to a clock with multiple alarms set to ensure timely progression through its phases: G1 (growth), S (DNA synthesis), G2 (preparation for mitosis), and M (mitosis). At each checkpoint, the cellular machinery assesses conditions before moving forward, like a meticulous inspector verifying the integrity of each step. Only if DNA is intact and the cell is ready will signals be sent to advance through the cycle.

In stark contrast, cancer cells hijack these controls, leading to relentless division regardless of damage or readiness. Mutations in genes that regulate the cell cycle, such as those coding for tumor suppressors or oncogenes, permanently disarm the checkpoints that ordinarily halt division when abnormalities are detected. This unchecked proliferation mirrors a clock with its alarms silenced; even as issues arise, time—represented by the cell cycle—marches on without intervention.

As regular cells meticulously coordinate division to maintain tissue health, cancer cells multiply rapidly and haphazardly, disrupting the balance and leading to tumors. Understanding the underpinnings of this process sheds light not only on the nature of cancer but also on potential therapeutic targets, as restoring control to the cell cycle could rein in malignant growth, much like rebooting a malfunctioning clock can reestablish order and function.

In the realm of the cellular lifecycle, certain proteins stand as guardians of regulation, ensuring each cell division is a measured, deliberate event. These include cyclins and cyclin-dependent kinases (CDKs), which act together like a time-sensitive lock and key system, allowing the cell cycle to progress only when conditions are ideal. CDK inhibitors serve as an additional checkpoint; if they detect DNA damage, they pause the cycle, buying time for repair, akin to an emergency brake when potential hazards are identified.

Yet, this system is not impervious to flaws. Mutations in DNA can give rise to faulty versions of these proteins, or even disable them entirely. Oncogenes, which are like the gas pedals of cell growth, can become stuck in the 'on' position due to mutations, driving cells to divide endlessly. Tumor suppressor genes, akin to brakes, can lose their ability to slow down or stop the cell cycle, removing the crucial failsafes that prevent unchecked growth.

Once these mutations establish a foothold, they trigger a cascade of events leading cells down the path of relentless division, marking the beginnings of a tumor. Cancerous cells often acquire the ability to shut down apoptotic processes – the cell's in-built self-destruct protocol – allowing them to survive and proliferate even when severely damaged or faulty.

Combatting this insurrection, modern medicine employs targeted therapies that are designed to specifically identify and thwart these rebellions at the molecular level. Some drugs focus on reactivating suppressed tumor suppressor genes; others introduce synthetic lethality, exploiting cancer cells' weakness to bring about their apoptosis, effectively hitting the 'reset' on the clock when it malfunctions beyond repair.

By dissecting these stages and mechanisms, it becomes possible to gain a precise understanding of the transition from a healthy cell to a cancerous state, offering insights into the development of treatments and perhaps, more critically, prevention strategies that might one day turn the tide in the ongoing battle against cancer.

Imagine the human genome as an extensive, intricate library of recipes, with each gene providing the instructions to whip up the proteins that keep our cells running smoothly. Genetic mutations are like typographical errors in these precious recipes—if you're lucky, a tiny typo might not change the dish at all, or it might just add an unexpected twist to the flavor. But occasionally, these errors fundamentally change the recipe, turning a batch of brownies into a savory bean casserole.

These genetic typos can range from a single misplaced letter to an entire paragraph inadvertently copied or deleted. Sometimes, the results are harmless; other times, they can be disastrous, leading to proteins that don't work properly or are made when they shouldn't be, much like adding way too much salt by mistake—it's not just the taste that's affected; it's the integrity of the whole dish.

In the body's case, such significant mix-ups can disrupt the delicate balance of cellular growth and repair, leading to diseases like cancer. It's as if one erroneous recipe starts a chain reaction in the kitchen, with pots and pans overflowing until the whole culinary process is in chaos. Understanding these errors not only sheds light on the quirks of biology but also guides us in crafting targeted treatments, like an experienced chef who knows just how to salvage a meal gone awry.

Let's take a deeper look at the intriguing world of genetic mutations, putting a microscope to their role in cellular functions and the onset of diseases such as cancer. Different types of genetic mutations include:

- **Point mutations**: These are small-scale changes affecting a single base pair within the DNA molecule. Consider it like a single typo where one letter is replaced by another, potentially changing a word's meaning entirely.

- **Insertions**: Here, extra base pairs are inserted into a DNA sequence. Imagine inserting a word into a sentence that doesn't belong, which can change the context or make it nonsensical.

- **Deletions**: Base pairs are deleted from the DNA sequence, akin to removing an essential word from a sentence, which might alter its meaning.

- **Chromosomal rearrangements**: Large segments of chromosomes are duplicated, deleted, inverted, or swapped with parts of other chromosomes. Think of it as taking entire paragraphs from one chapter of a book and mixing them with another, which can dramatically change the story's flow.

These mutations can lead to the production of defective proteins—or no protein at all—or they may cause a gene to be overactive. In a cell, this could mean creating a dysfunctional enzyme or switching on growth signals when they're not needed, akin to a light switch that's stuck in the 'on' position.

Normally, cells have a proofreading system to correct DNA errors— molecular spell checkers that fix typos to maintain the genetic integrity. But when this system fails, mutations persist, accumulating over time and potentially leading to diseases.

Particularly, oncogenes and tumor suppressor genes are critical in cell regulation. Oncogenes are like the gas pedal for cell growth, while tumor suppressor genes act as brakes. Mutations in these genes can cause the cellular car to accelerate out of control or lose its ability to stop, both leading paths towards cancer.

Genetic testing and screening are akin to quality control checks in a manufacturing process. They can identify mutations before they cause harm, allowing individuals to take preventive measures—like lifestyle changes or more frequent monitoring to catch and control a condition early on.

Finally, gene therapy holds the possibility of precisely fixing genetic errors, editing DNA like a meticulous copyeditor. By reverting harmful mutations, gene therapy seeks to restore normal function, offering hope in conditions where conventional treatments may fall short. This journey from pinpointing errors in DNA to effectively correcting them encompasses our growing understanding of heredity, disease, and the potential within our genes—to both create and cure.

Cancer's journey begins with initiation, where a cell acquires a mutation that nudges it toward uncontrolled growth. Think of this first stage as a switch flicking on, allowing the cell to bypass the usual growth regulations. During the promotion stage, this rogue cell benefits from additional genetic changes that promote its proliferation, much like a snowball gaining size and momentum rolling downhill.

Progression is the phase where a mass of abnormal cells, now called a tumor, establishes itself, developing its own blood vessels, a process known as angiogenesis, in order to sustain its growth. Imagine the tumor as a bustling construction site, setting up its own infrastructure to support the growing demands of its expanding population.

Finally, in metastasis, the most sinister phase, cancer cells embark on a journey to colonize new areas of the body. They break away from the original tumor, travel through the bloodstream or lymphatic system, and implant themselves in new tissues much like invasive plant species spreading far from their initial roots.

At each phase, the biological changes compound, not only increasing the number of mutated cells but also creating a cellular environment that supports the cancer's encroachment and survival. Understanding these stages in such detail illumines the complexity of cancer and underscores the need for multi-targeted approaches in its treatment, aiming to interrupt the progression and spread at various points in its path. It's a narrative of transformation, where each step presents both challenges and opportunities for intervention, providing invaluable insights into the creation of strategies to combat cancer's relentless advance.

Here is the breakdown of the critical biological and molecular events that chart the progression of cancer, shedding light on the transformation from a single mutated cell to a full-blown invasive disease:

- **Initiation:**
    - Mutations occur due to various factors, like chemical exposure or radiation, leading to DNA damage. These mutations can activate oncogenes or deactivate tumor suppressor genes.
    - Proto-oncogenes, when mutated, become oncogenes—genes that can promote cell division in the absence of normal growth signals, much like a car accelerator stuck in the 'drive' position, pushing cells towards proliferation.

- **Promotion:**
    - Proliferation can be accelerated by factors such as hormones and chronic inflammation, which provide a conducive environment for cancer cell growth.
    - The concept of clonal expansion comes into play, where mutated cells

with a growth advantage increase in number and dominate the cell population.

- Progression:
- Angiogenesis, the formation of new blood vessels, supplies the growing tumor with nutrients and oxygen, similar to building new roads for supplying a new city.
- Cellular adhesion and the extracellular matrix, the scaffold supporting cells, undergo changes, allowing the tumor to anchor itself more firmly and expand.

- Metastasis:
- Cancer cells acquire the ability to break away from the primary tumor, enter the blood or lymphatic systems, and travel to distant sites, where they can form new, secondary tumors.
- The microenvironment of these new sites can be either hostile or hospitable to the cancer cells, determining whether a secondary tumor will establish itself. Factors such as tissue-specific receptors and the presence of growth factors influence where metastases are likely to form.

The intricate series of changes from healthy cell regulation to the turmoil of cancer embodies a process where minute molecular alterations can have profound, body-wide consequences. By educating ourselves on these stages, their implications, and the biological cascade they trigger, a clearer understanding of the disease materializes, highlighting the importance of early detection and targeted therapies. This deep dive into the cellular rebellion not only acquaints us with the enemy within but also equips us with knowledge on how to counteract its advances.

The narratives of individuals like Steve Jobs, who faced pancreatic cancer, or Angelina Jolie, who took proactive measures against breast cancer, resonate not as mere tales of medical hardship but as human experiences intricately woven into the fabric of scientific discovery. Jobs' journey, battling a rare form of the disease, is like traversing an uncharted forest, facing unknown challenges under the canopy of modern medicine. His initial turn to alternative treatments exemplifies a navigation through the thicket of options available, a personal odyssey that many encounter.

Angelina Jolie's decision to undergo a preventive double mastectomy after discovering her BRCA1 gene mutation is akin to a skilled architect reinforcing the structure of a building long before the first signs of trouble. Her actions illuminate the power of genetic understanding and its potential in shaping healthcare decisions, much like a forecast prompts a farmer to shield crops before a storm.

As their stories unfold, we're reminded that beyond the clinical terms and complex mechanisms at play, there lies a personal narrative that echoes the fight against an internal uprising, a cellular rebellion that seeks to claim our very essence. Through their experiences, we gain a unique lens to view the science of cancer, where every breakthrough is a beacon of hope and each struggle a reminder of the ongoing battle within our biology.

Let's take a closer look at the intertwining of genetics and cancer risk by weaving through the real-life stories of impactful individuals. When discussing BRCA1, the gene associated with a higher risk for breast cancer, we can look to Angelina Jolie's proactive decision for a double mastectomy. Understanding that mutations in this gene can disrupt the normal repair of DNA, leading to more errors and, in turn, increased cancer risk, shows why Jolie's choice was informed and potentially life-saving. It's a decision that highlights the power of genetic awareness and preventative action, akin to renovating a building's foundation to withstand future earthquakes.

In the case of Steve Jobs, one contemplates the interplay between his specific condition and the broader landscape of pancreatic cancer. His disease was driven by mutations that caused cells in the pancreas to grow uncontrollably, a process similar to an assembly line that cannot be stopped, clutters the factory floor, and eventually halts all production.

These narratives provide insight into somatic mutations that arise from environmental and lifestyle factors and germline mutations that are passed through families. They unravel how different mutations prompt various approaches to treatment, much like personalized adjustments that must be made to a car based on model and use.

It all converges on a vital concept: targeted therapies mark a shift towards precision medicine, where treatments are customized to the genetic profile of an individual's cancerous cells. It's an approach that aims to fix the specific 'broken parts' rather than the 'entire machine,' offering a more effective and less invasive way to combat the disease. By extracting lessons from these personal stories, cancer genetics become less abstract, drawing a vivid picture of the battle between mutation and medicine. Through these dialogues, we understand that each stride in research and every informed health decision is another step towards mastery over one of humanity's most persistent adversaries.

The contemporary battlefield of cancer treatment is a terrain where traditional and innovative therapies coalesce. Surgery, the most time-honored approach, physically removes the cancerous growth, much like cutting away a section of damaged fabric to prevent fraying from spreading. Chemotherapy functions on a cellular level, introducing drugs that poison rapidly dividing cells, a scorched-earth strategy that targets the insurgents but also affects the land's healthy inhabitants. Radiation therapy is more precise, using high-energy particles to damage the DNA of cancerous cells, burning out the rebellion in specific locales with focused intensity.

Emerging targeted therapies represent a shift towards intelligent armament; they are akin to drones designed to seek and destroy specific molecular targets peculiar to cancer cells. This method offers the potential to minimize collateral damage by sparing healthy cells. Each modality in the cancer therapy arsenal comes with its own strengths and application criteria, and oftentimes, a combined arms approach bolsters the odds of success. These methods illustrate the versatility and evolution of oncological strategies, a testament to the ingenuity behind medical science's relentless pursuit to suppress the chaos of cellular mutiny.

Surgery is often the first-line strategy, beginning with a thorough diagnosis that might involve imaging like MRI or CT scans to map the extent of a tumor. Planning for surgery is meticulous, with the aim to remove the entire tumor while sparing as much healthy tissue as possible. Margins—or the boundaries of the removed tissue—are examined to ensure no cancerous cells are left at the edges. Postoperative recovery includes monitoring for complications and ensuring the patient's swift return to health.

Chemotherapy agents work because they target rapidly dividing cells—a hallmark of cancer. These agents can damage DNA or interfere with cell division, which, during the cell cycle's most vulnerable phases, potentially destroys cancerous cells. Combination therapies, using different drugs, exploit multiple weaknesses in cancer cells simultaneously, increasing the chances of treatment success while hoping to limit resistance.

Radiation therapy involves precise targeting of high-energy rays to damage the DNA of cancerous cells, leading to cell death or the inhibition of growth. Dosage and scheduling are carefully calculated to maximize tumor damage and minimize harm to healthy tissues, employing advanced techniques like intensity-modulated radiation therapy (IMRT) for improved precision.

The science of targeted therapies is a tale of matchmaking between drugs and genetic targets on cancer cells. It begins in the lab, identifying specific proteins or genes driving a cancer's growth. Researchers then develop or repurpose drugs to block these targets. These therapies are rigorously tested in clinical trials to ensure their efficacy and safety before becoming a part of standard treatment regimens.

Combining these treatments often follows a sequel from surgery to clear the bulk of the tumor, chemotherapy to address any remaining cells, and radiation to treat specific areas if needed. Targeted therapies may be woven throughout this sequence to exploit genetic vulnerabilities. Factors such as tumor type, stage, location, and overall patient health inform this multi-disciplinary approach, tailoring treatment plans to individual needs. The ultimate goal is to maximize the chances of a cure or prolong survival while maintaining the best quality of life for the patient.

The essence of the chapter hinges on the delicate balance of cellular regulation, a process intricately designed to maintain harmony within the body. When this balance is disrupted, it can set the stage for cancer, transforming an orderly cell community into a chaotic landscape. This disarray stems from genetic mutations that tilt the scales, pushing cells towards unchecked division and growth. However, the journey through this chapter reveals not despair, but the remarkable strides made in understanding and potentially reversing this anarchy.

At its core, the battle against cancer centers on deciphering and re-establishing order. Innovative treatments aim to target the very roots of cellular insubordination, whether by removing the rebelling factions through surgery, poisoning their supplies with chemotherapy, halting their advances with radiation, or outsmarting them with targeted therapies. Each strategy underscores the triumph of precision over brute force, of tailored approaches over one-size-fits-all remedies.

By delving into the specifics of these treatments, readers can appreciate the critical role of cellular governance and the continuous quest to refine our methods to reassert control. Each advance brings us one step closer to transforming the prognosis of cancer from a sentence of chaos to a challenge that can be met with hope and a well-stocked arsenal, turning the tides within this microcosmic battlefield.

CHAPTER 4: TYPES AND CLASSIFICATIONS OF CANCER

Chapter 4 swings open the doors to the vast and varied world of cancer types, each distinct in its origins, behavior, and impact on the human body. To navigate this complexity, a classification system acts as a compass, orienting patients and healthcare professionals alike in the landscape of oncology. This is where we begin to decode the language of cancer – sorting the multitude of cancers first by their starting locations, be it breast, lung, or skin, much like sorting books in a library by their genres.

The classification system extends beyond location; cancers are also categorized by how their cells appear under a microscope and behave – their histology and molecular structure. These classifications bear critical implications, guiding treatment choices and informing prognostices. Understanding this taxonomy is key to demystifying cancer, not just for those in white coats but for anyone eager to gain clarity on a disease that touches millions of lives. This chapter is a step towards that understanding – clear, concise, and aimed at illuminating the path through cancer's complexities.

At the forefront of categorizing cancer, two primary methods stand out: classification by the location of the cancer's origin and by the type of cell involved. The location refers to where in the body the cancer began, akin to pinpointing the birthplace of a river before it branches and spreads. A cancer's name often reflects this origin; for instance, lung cancer begins in the lungs, whereas breast cancer starts in the breast tissue.

In addition to location, recognizing the type of cell that has become cancerous is critical. This is where histology, the study of tissues under the microscope, provides insights as precise as a tailor examining the weave of fabric to determine its quality. It can differentiate between cancers that originate from muscle cells, which would be labeled sarcomas, from those that begin in glandular tissues, known as adenocarcinomas, through their unique cellular patterns.

Together, these categorization methods provide a framework for understanding how cancers are identified and addressed. Each category or subtype comes with distinct characteristics and, consequently, specific treatment approaches. Grasping these concepts is akin to understanding the roots and branches of a tree; it sets the groundwork for further exploration of how cancer affects the body and how medical science responds to each challenge. As this chapter unfolds, these foundations will be built upon, offering clarity and insight into the diverse and intricate world of cancer types.

Let's take a deeper look at the intricate framework used to classify cancer by location and cell type, and how this categorization critically informs the direction of treatment and prognosis. Common origins for cancers include organs such as the lungs, breast, prostate, and colon. Each location presents a distinctive environment, influencing how the cancer behaves and responds to treatment. Lung cancer, for example, due to the organ's exposure to inhaled substances, often requires aggressive therapies that may combine radiation, surgery, and systemic drugs, while breast cancer treatment can be tailored based on hormone receptor statuses.

Cancer can arise from various cell types, with epithelial cells leading to carcinomas, the most prevalent form of cancer. Connective tissues give rise to sarcomas, known for their occurrence in bones and muscles, whereas cancers like leukemias originate from the blood-forming cells, presenting with systemic symptoms and requiring treatments like chemotherapy or stem cell transplants.

Under the microscope, the histology of a tumor—whether cells appear well-differentiated and thus resembling their tissue of origin, or poorly differentiated and more abnormal—can indicate how aggressively the cancer is likely to behave. Well-differentiated cells might grow more slowly and respond better to localized treatments such as surgery, while poorly differentiated cells may need a combination of therapies due to their tendency to grow and spread more rapidly.

The implications of cancer's origins and histology on treatment are vast. Carcinomas, for instance, might be treated with hormone therapies if they are hormone receptor-positive, whereas sarcomas might not respond to these treatments and instead require chemotherapy. The prognosis hinges on these classifications as well; a well-differentiated carcinoma may have a more favorable prognosis than a poorly differentiated sarcoma, emphasizing the need for accurate classification to guide treatment plans.

By unpacking these classifications, patients and medical practitioners can work together to establish a prognosis, tailor treatment protocols, and set realistic expectations for the disease's trajectory. This detailed understanding underscores the personalized nature of cancer care, ensuring that each patient receives therapy that is attuned to the unique profile of their disease.

Imagine a diverse cast of characters in an epic drama, each with unique storylines and personalities that converge to create a dynamic narrative. This is akin to the diverse array of cancer types, where each form of the disease is characterized by distinct features and behaviors. Just as a knight, a healer, and a scholar play different roles in a story, skin cancer, leukemia, and breast cancer each have unique modes of progression, treatment responses, and impacts on the individual's health.

Skin cancer, the knight, is often armored by its visibility on the body's surface, allowing for early detection and intervention. Leukemia, the healer, circulates within the blood and requires systemic strategies to manage its widespread influence. Breast cancer, the scholar, may hold keys in its hormone receptor status that guide targeted and thoughtful treatment plans.

This ensemble of malignancies presents a medical theatre where understanding each character's traits—the aggressiveness of their actions, the terrain they inhabit, and the tools they are vulnerable to—not only informs treatment but crafts the course of a patient's journey. The interplay between these types yields insights as rich as the stories of a well-crafted play, casting a spotlight on both their individuality and their collective impact on the landscape of human health.

Here is how distinct cancer types are characterized, providing insight into their progression and treatment:

- **Skin Cancer: Visibility and Early Intervention**
- Regular skin checks and mole mapping allow for early detection of unusual changes, increasing the chances of catching skin cancer in its initial stages.
- Dermatoscope use during examinations provides a magnified view of

moles, aiding in the identification of malignant features.

- Biopsies of suspicious skin lesions confirm diagnoses and enable early treatment, often with minor surgical procedures that offer high success rates.

- Leukemia: Systemic Nature and Treatment Modalities

- Composed of various types, leukemia typically affects white blood cells like lymphocytes or myelocytes, disrupting normal blood function.

- Chemotherapy remains the primary treatment, aiming to destroy cancerous blood cells and allow for the recovery of bone marrow function.

- Bone marrow transplantation may follow to restore the production of healthy blood cells, serving as a potential curative approach, especially in severe cases.

- Breast Cancer: Hormone Receptor Status and Targeted Therapy

- Estrogen and progesterone receptors are proteins on breast cancer cells that, when present, drive the cancer's growth in response to these hormones.

- The hormone therapy drug tamoxifen blocks estrogen receptors, impeding the tumor's ability to use estrogen for fuel.

- Aromatase inhibitors lower estrogen levels in postmenopausal patients, thereby reducing the cancer's potential growth stimulus.

- Metastatic Propensity Across Cancer Types

- Breast cancer commonly spreads to the bones, liver, lungs, and brain, while colorectal cancer frequently metastasizes to the liver.

- A study of metastasis sites guides therapeutic strategies, highlighting the need for systemic treatments like chemotherapy, or targeted therapies for specific metastasis locations.

- Advanced stages often involve multimodal treatment approaches and palliative care to manage symptoms and improve life quality.

Understanding these details enhances the reader's grasp of the complexities of cancer, underscoring the tailored approach necessary in oncology, which factors in the unique behavior and characteristics of each type of cancer. With informed knowledge comes the power to navigate the challenging landscape of cancer treatment and care.

At the molecular heart of cancer lies a series of genetic mutations, altercations to the DNA sequence that can hijack a cell's normal functions. Like a series of typos in a critical instruction manual, these mutations can cause cells to grow and divide uncontrollably. Understanding these mutations paves the way for classifying and treating cancer.

To begin, there are oncogenes, akin to a car's accelerator stuck on 'go,' which promote cell division. Conversely, tumor suppressor genes act as the brakes, preventing uncontrolled growth. Mutations that activate oncogenes or deactivate tumor suppressor genes can effectively remove the checks and balances on cell proliferation, setting the stage for cancer.

The identification of these mutations is pivotal, as they inform targeted therapies—treatments that home in on specific altered genes or proteins driving the cancer's growth. Think of this as a precision-guided system designed to correct or counteract the effects of specific mutations. For example, the presence of a mutation in the gene HER2 in some breast cancer patients has led to the development of drugs like trastuzumab, which directly targets that gene product.

By breaking down cancer to its molecular underpinnings, one sees how personalized medicine can emerge. This approach allows treatments to be tailored, much like a suit adjusted to fit the individual measurements, ensuring the impact is direct and potent. Understanding these basic genomic concepts isn't just an academic exercise—it's central to the next generation of cancer care, empowering patients and their healthcare teams to confront cancer with targeted strategies.

The molecular landscape of cancer is as complex as it is diverse, with each cancer type bearing distinct genetic aberrations that drive their development. Here we dissect the intricacies of these mutations and the pursuit of precision medicine catered to these molecular defects:

- Genetic mutations span a spectrum, with point mutations representing a change in a single DNA building block, whereas insertions and deletions involve the addition or loss of small DNA segments. Chromosomal rearrangements, on the other hand, involve more extensive sections of chromosomes being duplicated, inverted, or translocated. Each of these mutations alters the genetic code and can activate oncogenes or disable tumor suppressor genes, leading to uncontrolled cell growth and cancer.

- Genomic sequencing, a cutting-edge methodical approach, scrutinizes cancer tissues to unearth these mutations. This technique reads the DNA sequence in cancer cells, highlighting variations that contribute to the cancer's behavior, much like a proofreader scans a document to spot errors that may alter its meaning.

- Targeted therapy is grounded in the action against these molecular errors. For instance, kinase inhibitors block specific enzymes involved in cell growth, while monoclonal antibodies are designed to latch onto and inhibit the function of specific proteins on cancer cells. It's a sophisticated armamentarium aimed directly at the biological underpinnings of a patient's cancer.

- The path from mutation identification to therapeutic intervention is meticulous. It begins with mapping the genetic alterations, progresses through designing drugs targeting those precise mutations, advances through the rigorous gauntlet of clinical trials to test efficacy and safety, and culminates with regulatory approval before becoming a standard treatment option—one tailored to the genetic intricacies of a person's cancer.

- Personalized medicine in oncology champions the potential for tailored treatment regimens that can lead to improved outcomes. However, challenges persist, including the cancer's ability to develop resistance to targeted drugs, and the daunting variability of genetic mutations across and within cancer types. The benefits of these therapies can be significant, but so is the necessity for ongoing research and innovation to overcome hurdles such as these.

Understanding these genetic determinants is paramount in crafting effective personalized treatment plans for cancer patients. This microscopic exploration of cancer's molecular characteristics arms individuals with a clearer understanding of how such complex information shapes the future of cancer treatment, illuminating the road to more precise and individualized care.

Picture this: staging cancer is like mapping out a journey for a notorious rumor spreading through a school. Initially, the rumor might be confined to a single classroom—akin to Stage I cancer, localized and contained. As whispers spread to corridors and other classrooms, the stages progress, with

Stage II and III representing wider dissemination within the school. By the time the rumor is rampant in the cafeteria, on the school bus, and at the local hangout spots, it's reached Stage IV, much like cancer spreading to distant organs, a process known as metastasis.

Meanwhile, grading cancer is like assessing the wildness of a rumor; some are close to the truth (well-differentiated), while others become wild tales barely resembling the original story (poorly differentiated). A low-grade cancer is more likely to resemble normal, healthy cells and grow slowly, much the way a accurate story loses steam. A high-grade cancer, on the other hand, looks very different from normal cells and has a propensity to grow quickly, akin to a wild, sensational rumor that takes on a life of its own.

These classification systems guide the 'treatment' plans for managing the rumor—whether a quick correction in a classroom will suffice or if a full-blown assembly is needed. In cancer care, understanding the stage and grade helps determine the best approach—localized surgery, radiation, chemotherapy, or a mix—tailored to halt the spread of the disease and restore health, much like quelling the chaos of schoolyard gossip.

Let's take a deeper look at the precision behind cancer staging and grading, cornerstones of cancer diagnosis that inform the path of treatment and prognosis.

- Starting with **staging**; cancer stages range from I to IV:
- **Stage I**: Here, cancer is typically small and contained within the organ it originated from.
- **Stage II and III**: These intermediate stages indicate larger tumor sizes and potential spread to nearby lymph nodes or tissues.
- **Stage IV**: Also known as metastatic cancer, this stage suggests the cancer has spread to distant parts of the body.

- Objective measurements for staging include:
- **Tumor size** (T): The physical dimensions of the tumor measured in centimeters or millimeters.
- **Lymph Node involvement** (N): Whether the cancer has spread to nearby lymph nodes and the extent of that spread.
- **Metastasis** (M): Whether the cancer has spread to other parts of the body.

- Moving to **grading**, it's a reflection of how much cancer cells look like healthy cells:
  - **Grade 1 (Well-differentized)**: Cancer cells resemble normal cells and tend to grow slowly.
  - **Grade 2 (Moderately differentiated)**: These cells don't look exactly like normal cells and could grow or spread at a moderate rate.
  - **Grade 3 (Poorly differentiated)** and **Grade 4 (Undifferentiated)**: Both represent cells that appear abnormal and tend to grow rapidly.

- Each **stage** and **grade** leads to a tailored treatment protocol:
  - **Early-stage (I and II)** cancers might be managed effectively with surgery alone.
  - **Stage III** may require a combination of surgery and radiation or chemotherapy, depending on lymph node involvement.
  - **Stage IV** typically calls for systemic therapies like chemotherapy, targeted therapy, or immunotherapy, in addition to local treatments as needed.

- The interplay between staging and grading significantly influences prognosis. Early stages with lower grades frequently have a better outlook, with higher chances of remission and long-term survival. In contrast, high-grade, advanced-stage cancers might require more aggressive treatment and have a guarded prognosis.

This understanding is paramount because it offers not just the blueprint for combating the disease but also the hope and expectation of what's to come. It turns the unwieldy nature of cancer into a readied outline where oncologists and patients can strategies and confront the illness with insight and preparation.

In the image of inspiring tales, Terry Fox's Marathon of Hope stitches a bold patch, embodying the human spirit's resilience against osteosarcoma, a type of bone cancer usually found in the legs. His journey mirrors an uphill marathon, not just of physical endurance but of an unyielding resolve, as he turned his personal struggle into a beacon of hope for cancer awareness and research. The crux of his story isn't just about the race against cancer; it encapsulates the collective journey of those touched by the disease, highlighting each stride towards understanding and combating various cancer types.

Chadwick Boseman, a hero on and off the screen, waged a covert battle with colon cancer, displaying strength akin to his onscreen character, the Black Panther. His quiet struggle cast a spotlight on the importance of early cancer screening and the silent progression of a disease that doesn't discriminate by age or stature. Boseman's narrative offers a poignant reminder of the stealthy nature of cancer and the crucial need for vigilance in early detection, much like a warrior sharpening his senses for signs of an unseen adversary.

Their legacies transcend their experiences, rendering not just tales of fortitude but painting a call for greater awareness and medical advancements in the fight against cancer's multifarious forms. It's this spirited confrontation against a seemingly inscrutable foe that continues to fuel the quest for knowledge—turning each personal battle into a source of shared strength and hope.

Cancer impacts millions of lives, yet the experience is deeply personal, as seen through the stories of individuals like Terry Fox and Chadwick Boseman. Terry Fox, a Canadian hero, thrust osteosarcoma, a bone cancer, into the global spotlight with his Marathon of Hope. His perseverance encapsulated the urgency for research and the universal desire for cures. Chadwick Boseman, renowned for his roles on screen, privately contended with colon cancer – a reminder that early detection and treatment are paramount in the narrative of cancer survival.

These stories anchor cancer's intangible breadth to tangible individual realities. They underscore the need for awareness, the importance of early screening, and the advancements in treatment that can arise from understanding and supporting those facing the disease. In recounting their experiences, a dialogue unfolds—not just about the medical battle but about the indomitable human spirit that fights it. This is the essence of the cancer narrative: a collection of individual stories that together inspire a collective stride towards hope and healing.

CHAPTER 5: THE CAUSES AND RISK FACTORS

Cancer arises not from a single source but as a culmination of factors that interlace in complex ways. Like threads weaving a image, genetic predispositions intertwine with lifestyle choices and environmental exposures to form the risk landscape of this multifaceted disease. Genetics can load the gun with mutations that may predispose individuals to cancer, while lifestyle factors like smoking or a sedentary life can pull the trigger. Meanwhile, environmental aspects, such as exposure to harmful sun rays or industrial chemicals, add layers of risk.

Genes like BRCA1 and BRCA2 are integral parts of the body's DNA repair system, much like quality control inspectors in a factory. When they function correctly, they help fix the mistakes in DNA that can lead to cancer. However, inheriting faulty versions of these genes—as can be checked through genetic testing—can significantly increase an individual's risk of developing breast and ovarian cancers, among others. This hereditary risk is like receiving a flawed blueprint that makes a building more susceptible to structural issues over time.

Angelina Jolie's case provides a compelling real-world example. After genetic testing revealed she had inherited a harmful BRCA1 mutation, she decided to undergo preventive surgeries, including a double mastectomy and an oophorectomy. This proactive approach, akin to renovating a building before critical issues arise, likely reduced her risk of developing cancer associated with the mutation.

Jolie's decision highlights how understanding genetic risk can inform personal healthcare choices. By taking charge of her health, she not only likely improved her own outcomes but also brought global attention to the importance of genetic testing and preventive care. Her choice serves as an example of how advanced knowledge of genetics can be a powerful tool in the fight against cancer, transforming potential risk into proactive prevention.

Looking deeper at the intricate roles of BRCA1 and BRCA2 genes and the profound impact their mutations have on our health and choices. These genes are akin to quality assurance in the body's cellular machinery, diligently correcting DNA errors and maintaining genomic stability. When BRCA genes function optimally, they produce proteins that help repair DNA breaks, ensuring cellular integrity and warding off cancerous changes.

However, mutations in these genes can derail this repair process. For instance, point mutations alter single nucleotides, the basic units of DNA, while larger deletions can remove entire chunks of the gene. Either way, the resulting protein might be rendered defective or wholly absent, leaving damaged DNA unaddressed and increasing cancer risk.

Genetic testing for BRCA mutations generally follows clinical guidelines that consider an individual's personal and family history of cancer. A blood or saliva sample can reveal the presence of mutations, guiding critical decisions in cancer prevention. When a mutation is identified, healthcare providers can personalize risk management strategies, advising increased surveillance, chemoprevention, or prophylactic surgeries.

Such surgeries, including mastectomy, or the removal of breast tissue, and oophorectomy, the removal of ovaries, drastically reduce the chances of developing associated cancers but come with significant considerations. Physically, they may entail long recovery periods and adjustment to changes in body image, while psychologically, they can impact an individual's sense of identity and future planning.

For those who test positive for BRCA mutations, implications extend to family planning and emotional health. Genetic counseling becomes invaluable, providing a clear understanding of the risk for future generations and the decision-making support needed to navigate these risks.

By meticulously dissecting the way BRCA genes affect our biological narrative, individuals are armed with empowering knowledge. With clarity on the nature of these mutations and the subsequent steps available, the path towards proactive health decisions becomes illuminated, transforming fear into actionable prevention.

Much like how our driving habits can affect the longevity and performance of a car, daily habits can have accumulating effects on our bodies, influencing long-term health outcomes. Consider how frequently revving a car's engine or skipping regular maintenance can lead to wear and tear that eventually requires significant repair. In a similar vein, the choices we make each day—be it smoking cigarettes, indulging in a nutrient-poor diet, or leading a sedentary lifestyle—compound over time, potentially accelerating the wear on our biological systems and increasing the risk for diseases like cancer.

Lance Armstrong's story serves as a poignant example. Renowned for his physical fitness and athletic achievements, Armstrong's lifestyle could not shield him from a cancer diagnosis. It stands as a testament to the complex interaction between lifestyle and genetics, and it underscores the fact that while we can manage risk factors, no one is immune to cancer. Armstrong's subsequent advocacy for cancer awareness and research brings to light the importance of recognizing the potential impact of our daily habits on our health and the necessity of proactive screening and prevention, much like routine check-ups and tune-ups that keep our vehicles—and by extension, our journeys—running smoothly.

Here is the breakdown of lifestyle factors and their complex interplay with cancer risk:

- **Tobacco Smoke and Cellular Biochemistry**
  - Carcinogens in Tobacco:
  - Includes chemicals like benzopyrene and nitrosamines.
  - These attach to the DNA in cells, causing mutations that can lead to cancer.
  - Cellular Impact:
  - Damages crucial genes that control cell growth and death, leading to uncontrolled cell proliferation.

- **Dietary Habits and Cancer Risk**
  - Processed Meats:
  - Often contain nitrites that can form N-nitroso compounds, proven to be carcinogenic.
  - High intake associated with colorectal cancer.
  - High-Fat Foods:
  - Can increase the production of hormones, notably estrogen, which is linked to breast cancer.
  - May lead to obesity, a known risk factor for various cancers.
  - Role of Inflammation and Oxidative Stress:

- Chronic inflammation can damage DNA over time.
- Oxidative stress can lead to changes in cell structure and gene expression patterns.

- Physical Inactivity's Effect on Hormonal Balance and Immunity
- Altered Hormonal Balance:
- Sedentary lifestyles can lead to hormone imbalances, including increased insulin and estrogen levels, promoting cancer growth.
- Immune Response:
- Regular exercise is known to boost the body's immune surveillance, which can detect and destroy cancer cells.

- Chronic Stress and Cancer Susceptibility
- Impact on DNA Repair:
- Long-term stress can suppress the body's ability to repair genetic damage.
- Stress Hormones:
- Cortisol, a hormone released in response to stress, can inhibit the effectiveness of the immune system.

- Sleep, Circadian Rhythm, and Cellular Health
- Sleep Disorders:
- Disrupted sleep patterns have been linked to higher rates of breast, colorectal, and prostate cancers.
- Circadian Rhythm Influence:
- The body's circadian clock regulates cell cycle and metabolism; disruptions can affect these processes and potentially lead to cancer.

Each lifestyle factor contributes a piece to the larger puzzle of cancer causation. While the relationships are complex, understanding them equips individuals with the capability to make informed decisions to potentially lower their risk. It's about making daily choices with an awareness of their long-term effects on our health, shaping not only our current well-being but our future vitality.

Within the vast array of substances that individuals might encounter on a daily basis, a number carry the notorious classification as carcinogens—agents with the capability to initiate cancer. Among these is asbestos, a fibrous mineral once widely used for its resilience to heat and corrosion.

When inhaled, asbestos fibers can embed themselves deep within the lung tissue, over time causing cellular damage that can lead to lung cancer, asbestosis, and mesothelioma, a malignancy of the lining of organs such as the lungs or abdomen.

The history of industrial workers' exposure to asbestos is a stark narrative that underscores the intersection of occupational hazards and health. For much of the 20th century, workers in construction, shipbuilding, and automotive industries frequently handled asbestos without protective measures or even awareness of its potential dangers. It wasn't until the latter part of the century that policies and regulations, driven by mounting evidence of health risks and public health advocacy, began to restrict asbestos use and implement safety protocols.

This reflection on past practices highlights not only the evolution of workplace health standards but also the importance of scientific vigilance in recognizing and mitigating risks associated with carcinogenic substances. Today, there is a continual effort to monitor and manage exposures, aiming to safeguard against the long-term health implications that industrial workers of previous generations faced without the benefit of such protective knowledge.

Industrial use of asbestos, a naturally occurring fibrous silicate mineral, began en masse in the late 19th century, with extensive mining starting in the early 20th century. It became a popular material in numerous industries for its durability, resistance to heat and chemicals, and insulating properties. Key industries included construction, automotive (for brake pads), shipbuilding, and the manufacturing of products like pipes and insulation.

The first inklings of asbestos-related health risks emerged as early as the 1920s and 1930s when medical studies began to associate asbestos exposure with lung problems in workers. However, it was not until the mid-20th century that asbestos's carcinogenic properties became widely accepted, with studies showing a strong link between asbestos and diseases such as asbestosis, lung cancer, and mesothelioma.

Regulatory action to protect workers was slow to follow these findings. In the United States, the Environmental Protection Agency (EPA) began banning certain asbestos products in the late 1970s and 1980s. Major strides were made with the implementation of the Asbestos Hazard Emergency Response Act (AHERA) in 1986 and the Asbestos School Hazard Abatement Reauthorization Act in 1990. Likewise, the Occupational Safety and Health Administration (OSHA) established regulations to reduce workplace asbestos levels.

Diagnosis of asbestos-related diseases usually involves imaging tests, such as X-rays or CT scans, to detect abnormalities in the lungs or pleura, and biopsies to confirm the presence of cancerous cells. The fibers can cause damage by physically irritating the tissue or by interfering with cellular repair processes, leading to mutations and cancer.

Current efforts to manage asbestos exposure include regular monitoring of airborne fibers in industries where asbestos is still present, stringent abatement protocols, and advanced personal protective equipment (PPE) for at-risk workers. For workers already affected, specialized treatment centers and legal provisions, such as compensation funds and litigation options, offer a semblance of recourse and support.

Understanding the history of asbestos and the evolution of workplace safety helps underscore the importance of vigilant occupational health practices. It reminds us that worker protection is an ever-evolving process, marked by continuous advancement in both regulation and technology.

Just as a computer virus can quietly infiltrate and compromise a system over time, chronic infections like Human Papillomavirus (HPV) can stealthily enter body cells and disrupt their normal functions, with the potential to cause changes that may lead to cancer. HPV, a viral saboteur, is particularly insidious because it can persist and go unnoticed for years, giving it ample time to launch a slow assault on the body's cellular machinery.

Certain strains of HPV have a predilection for causing cells in places like the cervix or the throat to rebel against their orderly nature, leading them down a path of unchecked growth that can culminate in cancer. This is what happened in the case of actor Michael Douglas, who bravely shared his experience with throat cancer linked to HPV. His candid revelation put a well-known face on a health issue, propelling greater awareness and understanding into the limelight.

Douglas's advocacy underscored the value of vaccination against HPV, which can act much like a firewall against the virus's cancer-causing properties. It's an example of how harnessing knowledge and taking preventative steps—much like updating antivirus software—can be potent tools in safeguarding well-being. By drawing attention to the connection between infections and cancer, he contributed to an ongoing narrative that stresses the importance of proactive health measures, turning personal trials into a collective call to action.

Let's take a deeper look at the intricate confrontation between the Human Papillomavirus (HPV) and the human body that can lead to cancer. The virus gains entry into the body typically through skin-to-skin contact, targeting the epithelial cells of the skin or mucous membranes. Once inside the cell, HPV integrates its DNA into the host's DNA, effectively hijacking the cell's machinery to replicate its genetic material alongside the cell's own.

Central to this hijacking are two HPV oncogenes, known as E6 and E7. These oncogenes produce proteins that bind to and degrade the body's tumor suppressor proteins, p53, and retinoblastoma protein (pRb), respectively. Under normal circumstances, p53 and pRb help prevent uncontrolled cell growth, but when their function is disrupted by E6 and E7, cells begin to divide unchecked. This can lead to the accumulation of mutations and, eventually, cancer.

High-risk HPV strains have been closely associated with various types of cancer, the most common being cervical cancer. They also play significant roles in the development of oropharyngeal and anal cancers, among others, highlighting the virus's broad oncogenic potential.

Various vaccines, such as Gardasil and Cervarix, have been developed to provide immunity against key cancer-causing strains of HPV. These vaccines work by introducing harmless viral particles that stimulate the immune system without causing infection, thereby pre-empting the virus's attack. To be most effective, vaccination is recommended before individuals become sexually active, with schedules varying based on age and immunization history.

Public figures like Michael Douglas have been paramount in educating the public on the connection between HPV and cancer, turning their personal health battles into catalysts for change. Douglas's openness about his throat cancer, for example, has not only spread knowledge but has also helped de-stigmatize the disease and encourage vaccination, influencing health policy toward broader adoption of preventative measures.

Understanding the mechanics of HPV's role in cancer development emphasizes the importance of vaccination as a formidable defense strategy. By demystifying how this virus operates and highlighting successful prevention efforts, one can appreciate the preventive power individuals wield when armed with the right knowledge and tools.

As our body's internal clock ticks forward, so does the natural aging process, marking not just the passage of time but also altering our biological landscape, accompanied by a shift in hormonal ecosystems—both of which are integral to cancer risk. With age, cells accumulate genetic alterations due to a decline in the body's repair mechanisms, somewhat akin to the fading precision of an aging clock. This gradual build-up of genetic wear and tear increases the likelihood of developing cancer since the body's ability to correct cellular defects diminishes, much as an old car struggles to pass an emissions test.

Hormonal fluctuations throughout life, particularly those associated with the reproductive cycle, also play a role. Estrogen and progesterone, for example, influence cell division in breast and uterine tissues. An imbalance or prolonged exposure to these hormones can heighten the risk of cancers developing in these areas, in the same way that an erratic climate can stress an ecosystem.

When considering the broader population demographics, it's apparent that certain cancer risks are more prevalent in older adults; this echoes the increasing incidence of cancer with advancing age. Hormone-related cancer risks, while influenced by age, are also shaped by an individual's life events, such as age at menstruation onset and menopause. It's a complex matrix, where timing and duration of hormonal exposure interlock with changes wrought by natural aging.

Detailing these factors illuminates how carefully orchestrated systems can sway under the pressures of time and hormonal shifts, providing crucial insight into why some cancers become more probable as one traverses through life's stages. This understanding reinforces the critical nature of age and hormone-specific screening protocols and lifestyle adjustments that can aid in anticipating and managing these varied cancer risks.

Cancers don't arise in isolation; instead, they result from a complex interplay of genetic, lifestyle, and environmental risk factors. These factors can collaborate in a cumulative fashion, often magnifying each other's effects to increase the likelihood of disease. For instance, the combined impact of smoking and asbestos exposure far exceeds their individual risks, underscoring the concept that when it comes to cancer, the whole risk can be greater than the sum of its parts.

Early detection and preventative measures stand out as the vanguard in the fight against cancer. Screenings such as mammograms and colonoscopies can catch cancers in their infancy, when they're most treatable, similar to how smoke detectors can alert us to a fire before it spreads. Vaccinations, like the HPV vaccine, act as a pre-emptive strike to ward off initial infections that could otherwise lead to cancer.

In essence, understanding these overlapping risks is paramount to crafting effective prevention strategies and emphasizing the role of regular health screenings. By taking timely action and utilizing available preventative measures, individuals can significantly reduce their cancer risks and improve their chances of successful treatment, should cancer arise. This proactive approach to health can empower people to take control of their wellbeing and maintain their quality of life.

CHAPTER 6: CANCER DETECTION AND DIAGNOSIS

Stepping into the nuanced world of cancer detection and diagnosis is akin to entering a realm where precision, innovation, and keen insight converge. The strides made in medical technologies and methodologies are nothing short of remarkable—a testament to human ingenuity and relentless pursuit of understanding. With novel imaging techniques that unveil the smallest tumor details, to advanced genetic testing that pinpoints specific cancer markers, this world is constantly evolving, pushing the boundaries of what's possible in medicine. Each new development enhances the ability to not only detect cancer earlier but to tailor treatments in a way that was once only imagined. This chapter invites you to look at this progressive landscape, illuminating the advancements that have transformed a once daunting pursuit into a path lined with hope and clarity.

Cancer screening is a proactive search for cancer before a person exhibits any symptoms, a preventive check akin to a mechanic inspecting a car before a long trip. The goal is to find and treat cancer early when it's most manageable. Mammograms, for example, are recommended annually or every two years for women over the age of 40 to detect breast cancer. Colonoscopy, advised to start at age 50 and repeated every 10 years, is the flashlight that illuminates the colon, seeking out growths that could become colorectal cancer.

Adherence to these recommended frequencies is determined by risk factors such as family history, genetics, and lifestyle. If risk is higher, the screenings might be as frequent as a car in rough conditions needing more frequent service. Following medical screening recommendations is crucial—it's the regular maintenance that can prevent a manageable issue from becoming a full-blown problem. Compliance ensures that any cancer is caught as early as possible, when the toolbox for treatment is full and the likelihood of a favorable outcome is at its highest.

In this methodical surveillance of one's health, each screening acts as a sentinel, guarding against the hidden progression of cancer. Amidst the complexity of risk factors and medical guidelines, the essence holds clear: early detection can save lives, making compliance with screening recommendations not just a suggestion, but a critical component of modern preventive healthcare.

Let's take a detailed look at the meticulous processes involved in cancer screening and the actions that follow a positive result. Mammography and colonoscopy are two mainstay screening procedures. Mammography uses low-energy X-rays to examine breast tissue, effectively seeking out distortions and microcalcifications that could signal the onset of cancer. Conversely, colonoscopy involves the insertion of a long, flexible tube with a camera, called a colonoscope, into the rectum to visually examine the entire colon for polyps or abnormal growths that could turn cancerous.

To personalize the frequency and type of screening, medical professionals assess personal risk factors. A detailed family health history may reveal a hereditary pattern of cancer, prompting earlier and more rigorous screening. Genetic testing can identify individual vulnerabilities, such as mutations in the BRCA1 or BRCA2 genes, that significantly raise the risk of developing certain cancers. Additionally, lifestyle factors like smoking or occupational exposure to carcinogens could warrant an adjusted screening schedule.

When a screening flags abnormal results, a structured protocol is followed. Confirmatory tests such as diagnostic mammograms, ultrasound, or biopsy for breast screening, and polypectomy or tissue biopsy in the case of colonoscopy, might be employed. These subsequent tests aim to conclusively determine the presence and nature of cancer. With the results in hand, the patient and healthcare provider engage in a thorough discussion to look at treatment options, considering the specifics of diagnosis, overall health, personal preferences, and in some cases, second opinions.

Through this sequence of tailored screenings and informed/action-driven responses to positive results, the complex undertaking of monitoring for cancer becomes a navigable path, ensuring that individuals receive the appropriate level of care with ample opportunity for timely intervention.

Biopsies stand as a crucial step in cancer diagnosis, akin to extracting a core sample to understand the composition of the earth. There are several types of biopsies, each tailored to different circumstances. A needle biopsy uses a fine needle to extract cells from tumors or fluid from a suspicious area, while a core biopsy involves a larger needle to remove a cylinder of tissue for more comprehensive analysis. For areas that are harder to reach, an endoscopic biopsy uses a flexible tube with a camera to visualize internal organs and retrieve tissue.

Once the sample is secured, it travels to the pathology lab, where it's processed and examined under a microscope. A pathologist scrutinizes the cells' appearance and behavior to ascertain if they're benign or malignant, their aggressiveness, and to pinpoint their origin. This micro-level interrogation can reveal the cancer's subtype, informing which medical treatments may be the most effective.

Understanding biopsy types and their execution aids in grasping how medicine draws from this intimate look at body tissues to craft an accurate diagnosis. This deep dive into the biopsy's role demonstrates the intricacy and precision of cancer diagnostics, offering a window into the microscopic world where cellular secrets unveil the larger narrative of an individual's health.

When selecting the most fitting biopsy, physicians consider the tumor's accessibility, size, and location, along with the patient's overall health. A needle biopsy is often used when the concern is accessible and less invasive methods are preferred, such as in the breast or thyroid. For larger or deeper concerns, a core biopsy can provide a more substantial tissue sample, ideal for prostate or lung lesions.

The technique of a needle biopsy involves numbing the area with local anesthesia and guiding a fine needle to the site to retrieve cells or fluid. In a core biopsy, a larger needle, often aided by imaging technology like ultrasound or MRI, extracts a cylinder of tissue, providing a wider snapshot of cellular architecture. Endoscopic biopsies require sedation as a flexible tube with a camera and biopsy tools is navigated to the target site, often within the digestive tract or lungs, to obtain a sample.

Once the sample reaches the pathology lab, precise and detailed work unfolds. Tissue samples are preserved, sliced thinly, and placed onto slides. Special dyes, or stains, are applied to reveal cellular components under a microscope. These staining techniques, such as the Hematoxylin and Eosin (H&E) stain, highlight the detailed structure of cells, making abnormalities more evident.

Pathologists examine these slides to detect malignancy, classifying the cells based on their behaviors and appearance. Combining biopsy results with imaging findings and blood tests, a full picture of the patient's condition emerges, influencing treatment planning. Should atypical cells be detected, further genetic tests may be warranted to uncover specific mutations, informing targeted therapies tailored to the patient's unique cancer profile.

In navigating the biopsy process, the goal is always clarity—a focused approach that translates minute cellular insights into a map for patient care, from discovery right through to recovery.

Imagine you've lost a precious earring in a dark room; your initial method might be to use a flashlight to sweep the area. This flashlight is like an X-ray, which can shine through the body, casting shadows of denser materials like bone, and potentially reveal the outline of a tumor. Now, suppose you need a more detailed search—switch on the overhead light and use a magnifying glass. This is similar to an MRI scan, which offers a detailed, magnified view of the body's interior landscape, using a strong magnetic field and radio waves to illuminate fine details that aren't visible with standard light.

But what if the earring is hidden within something, like the lining of a pocket? That's when you'd use your sense of touch to feel for its shape and texture through the fabric. In the medical world, this tactile exploration is akin to a CT scan, where multiple X-ray images taken from different angles are compiled to create cross-sectional views—a tactile depiction, of sorts, granting the ability to perceive the hidden nooks of the body, detect irregularities, and understand their exact location and size.

These imaging technologies form an integral part of the diagnostic process, much like various search techniques help uncover lost items. They allow doctors to non-invasively survey the mysterious inner workings of the body, revealing the presence of unwelcome intruders like tumors, and serve as the compass for navigating the journey of treatment planning. With every advance in imaging, the curtain veils the body's innermost secrets a little more, empowering physicians to diagnose with confidence and to tailor therapies that are as precise and personalized as the approach one might take in retrieving that missing earring from just the right angle.

Here is the breakdown of the attributes and advantages of X-ray, MRI, and CT scan technologies:

- **X-ray Imaging:**
  - X-ray Radiation Physics:
    - Uses electromagnetic radiation to visualize internal structures.
    - Tissues of different densities (e.g., bone vs. muscle) absorb X-rays at different rates, creating a contrast image.
  - Machine Setup:
    - Consists of an X-ray tube that emits radiation and a detector on the opposite side that captures passing rays.
    - Images captured are black and white, with denser materials appearing whiter.

- **MRI Technology:**
  - Components of an MRI Machine:
    - Superconducting magnets create a powerful magnetic field.
    - Coils send and receive radio waves, which influence the atoms in the body.
  - Image Creation Insight:
    - Radio waves alter the alignment of hydrogen atoms, which emit signals when returning to their original alignment.
    - These signals are processed to create detailed images of soft tissues.

- **CT Scan Mechanics:**
  - Rotating X-ray Beams:
    - A circular gantry rotates around the patient, emitting X-ray beams from multiple angles.
    - Offers a three-dimensional view by compiling many two-dimensional images.
  - Computational Processing:
    - Advanced algorithms transform the 2D X-ray data into cross-sectional 3D images.
    - Allows for the visualization of soft tissue, blood vessels, and bones in the same image.

Each modality's technology dictates its diagnostic use; X-rays for broad screening and bone injuries, MRI for detailed soft tissue analysis, and CT scans for a rapid, comprehensive body scan often used in emergency settings. The selection of which imaging method is suitable for a patient depends on the medical indication, with each providing a unique perspective of the body's internal structure, subsequently guiding diagnosis and informing treatment plans.

Molecular testing is akin to a detective closely examining clues at a crime scene to understand the event's specifics. These tests analyze cancer at a molecular level, delving into genes, proteins, and other markers to uncover the disease's intricacies. Tests can range from identifying specific mutations within DNA—like the presence of BRCA mutations that heighten breast cancer risk—to examining the levels of particular proteins that may influence cancer growth.

A fundamental principle behind molecular testing is to personalize cancer treatment. By pinpointing a tumor's unique genetic and molecular profile, treatments can be tailored to target the cancer cells effectively and efficiently, much like a locksmith crafting a key to fit a particular lock. For instance, understanding whether a breast cancer tumor expresses hormone receptors directs the use of hormone-blocking therapies in treatment.

The scientific foundation of these tests rests on biomarkers, biological substances that act as indicators of a particular biological state or condition. The presence, absence, or mutation of these biomarkers provides valuable insights into how the cancer will behave and respond to treatment. By examining the cells at this granular level, physicians can choose a treatment strategy that aligns precisely with the patient's cancer type, improving treatment efficacy and often reducing side effects.

In the evolving landscape of oncology, molecular testing illuminates a path to decision-making that is informed by the very building blocks of life, offering hope through precision medicine. These insights bridge the gap between a universal approach to cancer care and one that is as unique as the individuals receiving treatment.

Let's take a detailed tour through the landscape of molecular tests that are revolutionizing cancer treatment and underlying patient-specific therapeutic decisions.

- **Genomic Sequencing:** This process involves decoding the DNA sequence to identify mutations that could drive cancer. Once these mutations are identified, targeted therapies can be employed. These therapies are designed to specifically disrupt the molecular mechanisms that the mutated genes use to promote cancer growth. For instance, if genomic sequencing detects the presence of a mutation in the EGFR gene in lung cancer, a bespoke inhibitor drug designed to block that specific gene's pathway can be used.

- **Proteomic Analysis:** Proteomics involves mapping the complete set of proteins expressed by the genome, known as the proteome, which can vary from one cancer cell to another. Understanding which proteins are overexpressed in a tumor can lead to the development of drugs that target those specific proteins, akin to silencing an overly loud instrument in an orchestra to restore harmony.

- **Pharmacogenomics:** The principle here is that slight variations in an individual's genetic makeup can affect their response to certain drugs. By assessing these genetic variations, a more personalized and effective drug regimen can be tailored to the individual. For example, the identification of specific gene variants can predict whether certain drugs can be metabolized effectively, avoiding treatments that might be ineffective or cause adverse reactions.

- **Tumor Marker Tests:** These tests measure substances, often proteins, produced by cancer cells or by the body in response to cancer. They serve as beacons of the tumor's presence or proliferation and can indicate how well the body is responding to treatment. Elevated levels of certain markers, like CA-125 in ovarian cancer or PSA in prostate cancer, can flag the need for a treatment change or indicate a relapse.

Each molecular test plays a pivotal part in the precision medicine symphony, with every measure bringing us closer to therapies that are as individual as fingerprints. By peering deep into the genetic and molecular underpinnings of each patient's tumor, these tests craft a roadmap for the journey from diagnosis to remission. In essence, this is the science of hope, rendered with precision for the art of healing.

The TNM staging system and the grading of tumors are fundamental components in the landscape of cancer care, providing a framework to gauge the severity of an individual's cancer and inform their unique treatment path. TNM, an acronym for Tumor, Nodes, Metastasis, breaks down the complex reality of cancer into an understandable code.

- 'T' details the size and extent of the primary tumor, ranging from T1, for small, localized tumors, to T4, for large tumors or those spreading into nearby structures.
- 'N' refers to the involvement of lymph nodes. N0 denotes no lymph node involvement, while higher numbers indicate increasing involvement.
- 'M' stands for metastasis. An M0 score signifies that cancer has not spread to other parts of the body, whereas M1 indicates the presence of distant metastasis.

Grading, on the other hand, scrutinizes the appearance of cancer cells under a microscope, assessing how much they differ from healthy cells. Low-grade tumors resemble normal cells and tend to grow slowly, whereas high-grade tumors look significantly different and suggest a more aggressive growth pattern.

Therapeutic approaches pivot on these classifications: for a low-stage, low-grade cancer, surgery might suffice, whereas high-stage, high-grade cancers could necessitate a robust combination of radiation, chemotherapy, and possibly targeted therapies. These systems guide decisions every step of the way, from the calibration of treatment intensities to the frequency of monitoring post-therapy. In essence, understanding TNM staging and tumor grading equips both the patient and the medical team with the insights needed to chart a treatment course that's as precise as navigating by the stars – aiming for the best possible prognosis with the information at hand.

In clinical practice, the TNM staging and tumor grading systems are meticulously applied tools used to characterize and plan the treatment of cancer. Categorizing the 'T' aspect of the TNM system begins with diagnostic tests such as MRI, CT scans, or ultrasounds, which map a tumor's size and exact location. Standardized guidelines, like those from the AJCC's Cancer Staging Manual, provide universal tumor measurement protocols for consistency across cases.

Determining 'N' status, which relates to lymph node involvement, often entails imaging techniques to highlight enlarged nodes suggestive of cancer spread. Surgical lymph node dissection or sentinel node biopsy may follow, where nodes are surgically removed and examined for the presence of cancer cells, verifying the imaging indications.

Metastasis ('M') detection involves a combination of imaging studies, such as bone scans for skeletal metastasis or PET scans that can detect cancer activity throughout the body. A biopsy may confirm the metastatic nature of a particular lesion identified on imaging.

Tumor grading is a microscopic venture where pathologists analyze tissue samples, comparing cancer cells' appearance to that of their healthy counterparts. Grades range from well-differentiated (Grade 1), which closely resemble normal cells, to poorly differentiated (Grade 3 or 4), indicating a more aggressive cancer. This cellular detail reveals the tumor's likely behavior and growth rate.

Integrating the TNM staging with tumor grading, clinicians craft a patient's treatment plan. For localized, low-grade cancer, surgery may be the mainstay. More advanced or aggressive cancers could require a multimodal approach involving surgery, radiation, and possibly chemotherapy or targeted agents, depending on the tumor's specific genetic makeup.

These staging and grading systems provide a narrative for each patient's cancer, guiding the clinical team in choosing a treatment path that is as individualized as the patient themselves. The level of precision and detail in this process facilitates a targeted assault on cancer while preserving as much normal tissue and function as possible.

A pathology report is a document that turns the findings of a microscopic battlefield into a clear narrative, telling the story of a patient's tissue sample. Key components of this report include the diagnosis section, which states the type of cells found and whether they are benign or malignant. It's the ultimate verdict in the search for cancer.

The report will often include the size and location of the tumor, described in centimeters, giving a measure of its physical footprint. Other critical points are the margins, which are the edges of the sample: clear margins mean no cancer cells at the edge, whereas positive margins suggest some cells have been left behind.

Further layers of detail are added with information like lymphovascular invasion, which indicates whether cancer has reached the bloodstream or lymph system, potentially acting as a highway for cancer to travel elsewhere in the body.

Additionally, the report may provide grading of the tumor cells, revealing how much they resemble normal cells, which helps forecast the speed at which the cancer might grow or spread.

Lastly, crucial molecular details can pinpoint genetic markers that may predict how the cancer will behave or respond to certain drugs, ushering in targeted therapies to the fore of the treatment plan, tailored to the very DNA of the cells in question.

Interpreting these elements enables the medical team to strategize the next steps, whether that means surgery, radiation, or chemotherapy. A pathology report is like a commander's briefing in the fight against cancer, with every piece of intel guiding the tactics for victory in a personal war for health.

Here is the breakdown of the detailed components of a pathology report and their roles in cancer diagnosis and treatment planning:

- **Diagnostic Statement:**
    - Criteria for Classification:
    - Cells are classified as benign or malignant based on cellular structure, growth patterns, and invasion into surrounding tissues.
    - Benign cells resemble normal cells and are often non-threatening, while malignant cells differ greatly from normal and can spread, signaling cancer.
    - Impact on Treatment:
    - Benign findings may require minimal to no treatment, whereas malignant diagnoses typically lead to a more aggressive approach, including surgery, chemotherapy, or radiation.

- **Tumor Size and Location:**
  - Measurement Methods:
  - Utilize imaging studies such as MRI or CT scans and direct measures during surgery to determine the tumor's dimensions.
  - Influence on Treatment Options:
  - Location affects surgical resectability and whether adjacent organs are at risk. Size can influence the extent of surgery or radiation therapy needed.

- **Margin Status:**
  - Definitions:
  - Clear margins indicate that no cancer cells are found at the outer edge of the tissue removed, suggesting complete tumor removal.
  - Positive margins indicate that cancer cells are present at the edge, suggesting that some cancer remains.
  - Post-Surgery Treatment Considerations:
  - Positive margins often necessitate additional treatment, such as re-excision surgery or adjuvant therapies.

- **Lymphovascular Invasion:**
  - Significance of Detection:
  - The presence of cancer in lymph or blood vessels increases the risk of spread to other body parts.
  - Prognosis and Systemic Therapy:
  - It may prompt systemic therapy like chemotherapy to address potential metastasis and improve long-term prognosis.

- **Cellular Grading:**
  - Tumor Grading Scale:
  - Grades range from 1 to 4; higher numbers indicate more abnormal-looking cells under a microscope, generally associated with a more aggressive cancer.
  - Implications for Treatment Modality Choice:
  - High-grade tumors may require more aggressive treatment, while low-grade tumors might be managed with a conservative approach.

- **Molecular Details:**
  - Identification of Mutations:
  - Tests for mutations, such as EGFR, ALK, and BRCA, which are

linked to specific targeted therapies.
  - Treatment Personalization:
  - Knowing the cancer's genetic mutations allows for tailored therapies that specifically target the molecular pathways involved in the cancer's growth and survival.

Working through a pathology report involves piecing together a puzzle that reveals not only the nature of the tumor but also outlines the battlefield for treatment strategies. Crucial to both prognosis and planning, this report provides a scientific compass guiding the direction of a patient's cancer treatment journey.

High-profile figures have harnessed their public platforms to become flag-bearers in the realm of cancer detection advocacy. When celebrities like Angelina Jolie candidly disclosed her BRCA gene mutation and subsequent preventive double mastectomy, the ripple effect was immediate and substantial. Coined as the 'Angelina Jolie effect,' there was a marked surge in women seeking genetic testing for breast cancer risk, a testament to the power of visibility on decision-making around health.

Similarly, Katie Couric's on-air colonoscopy following the death of her husband from colorectal cancer prompted a noticeable spike in screenings. Termed the 'Couric effect,' it underscored how impactful public discourse can be in demystifying medical procedures and encouraging proactive health measures.

These narratives, amplified by their reach and influence, have converted personal struggles into catalysts for public health action. Increased screenings and a bolstered awareness around the critical need for early cancer detection serve as enduring testaments to the legacy of their advocacy. By laying bare their journeys with transparency and poise, these individuals have redefined the intersection of public life and public health, underscoring the invaluable role of advocacy in the fight against cancer.

Cancer screening stands as a critical line of defense in the pursuit of health and longevity, and embracing this tool is a powerful step individuals can take towards safeguarding their well-being. The advancement of diagnostic technologies continues to revolutionize how early and accurately cancer can be detected. Innovations such as liquid biopsies, which search for cancer DNA in blood, and artificial intelligence in image analysis, exemplify the forward march of medical capabilities that enhance early detection and intervention.

When individuals engage with regular screening practices, they contribute to the vital culture of prevention. Leveraging cutting-edge diagnostics, health professionals can offer more personalized and timely treatment, improving the odds of favorable outcomes. Each technological stride empowers not only the clinicians but also individuals, who can take charge with informed decisions about their health.

By remaining vigilant and participating in routine screenings, individuals harness these advancements for their benefit, ensuring the narrative of their health does not surrender to chance. As the landscape of cancer screening and diagnostics broadens, it anchors the collective resolve to triumph over cancer, one early detection at a time.

CHAPTER 7: TREATING CANCER

In the changing world of oncology, the phrase 'one size fits all' is becoming a relic of the past. Chapter 7: Treating Cancer delves into the intricate matrix of current treatment modalities, each meticulously painted to fit the unique canvas of the patient's condition. The era of personalized medicine is ushering in a revolution, where the patient's voice and genetic blueprint steer the course of therapy. This chapter uncovers the collaborative nature of treatment decisions, spotlighting the empowered role patients play alongside healthcare professionals in confronting cancer. From the decision-making process that tailors therapy to the micro-level of genes, to the broad brushstrokes of combining treatment methods, understand the meticulous art and science behind each chosen path. These are not just clinical decisions; they are personal choices, with every option weighed and every protocol finely tuned to usher in the best possible outcomes for those at the heart of the treatment—the patients themselves.

The primary methods employed in the battle against cancer—surgery, chemotherapy, and radiation therapy—serve as the pillars of conventional treatment. Surgery, often the first option, involves physically removing the tumor. Much like cutting out a piece of rotten wood to prevent the spread of decay, surgeons excise cancerous tissue. This approach is most effective when the tumor is localized and hasn't spread, aiming for complete removal and sometimes followed by other treatments to eliminate any remaining cells.

Chemotherapy circulates anti-cancer drugs throughout the body, targeting rapidly dividing cells, a characteristic of malignancy. Picture a lawn being sprayed with weed-killer; it's not just the weeds that are affected but the entire area. Similarly, chemotherapy can impact healthy, fast-growing cells, leading to side effects like hair loss and fatigue. It's often utilized when cancer has metastasized or as a precaution to catch microscopic cells that may have escaped the primary tumor site.

Radiation therapy, on the other hand, harnesses high-energy particles to damage DNA within cancer cells, stifling their ability to multiply. Think of it as using a laser to precisely burn away a tumor, with a focus on minimizing damage to the surrounding, healthy tissue. Radiation therapy is regularly recommended alongside surgery to target any residual cancer cells, or it can be used as a standalone treatment in cases where surgery isn't an option.

Understanding when and why each of these treatment options is recommended hinges on a detailed analysis of the type, location, and stage of cancer, as well as the patient's overall health and personal preferences. Each treatment carries its own set of risks and benefits, and often, the best approach combines multiple methods, strategically sequenced for maximum effect—like a tailored multi-course meal designed to provide a satisfying, well-rounded experience.

The decision-making process is as critical as the treatments themselves, with ongoing research continually refining these modalities. A clear grasp of these options equips patients and healthcare providers to navigate the complexities of cancer treatment, not just with hope, but with informed confidence.

Let's take a closer look at the specific techniques and strategies that form the core of cancer treatments, particularly focusing on the variety of surgical methods, the spectrum of chemotherapy drugs, and the range of radiation therapies available.

Surgical options have evolved from traditional open surgeries to minimally invasive procedures. Open surgeries involve large incisions, providing surgeons direct access to the tumor. Minimally invasive surgeries, like laparoscopy, require smaller incisions, use specialized equipment, and typically result in quicker recoveries. Robotic-assisted surgeries represent the forefront of precision, with robots offering surgeons enhanced dexterity and control. These methods are chosen based on the tumor's location, size, the patient's overall health, and the goal of achieving clear margins.

Chemotherapy drugs can be categorized based on their mechanism of action. Alkylating agents, for example, work by interfering with DNA replication, preventing cancer cells from multiplying. Antimetabolites mimic normal substances within the cell, disrupting function and growth. Antimitotics inhibit cell division, targeting the microtubules. These drugs are selected based on the type of cancer, its growth rate, and how it responds to particular drug actions.

Radiation therapy techniques are characterized by their delivery method and the precision with which they target tumors. External beam radiation therapy (EBRT) projects radiation from outside the body, focusing on the tumor site. Brachytherapy, however, places radioactive sources directly into or near the tumor, delivering high doses with minimal impact on surrounding tissues. Stereotactic radiosurgery, despite the name, is a non-surgical procedure that delivers precise, high-dose radiation to a small, focused area—often used for brain tumors.

Each modality requires careful consideration of the cancer's specific traits—its cellular makeup, how aggressively it grows, and its stage. By understanding the subtleties of how these treatments impact cancer cells, medical teams can craft personalized treatment protocols that address the unique parameters of an individual's cancer, enhancing the chances of successful outcomes and, where possible, preserving healthy tissue. This nuanced approach to treatment selection underpins the patient-centered care that characterizes modern oncology.

Surgery in cancer treatment serves as a cornerstone, with several kinds woven into the fabric of oncological care. Excisional surgeries, where the entire tumor is removed, act as a precise strike, aiming to excise the cancerous tissue in its entirety, not unlike removing a problematic tree from one's backyard to halt the spread of disease. This is often the preferred option when the tumor is localized and the objective is curative.

When the tumor is large or has spread and complete removal isn't feasible, debulking surgery comes into play. The goal here is to reduce the tumor's size, akin to pruning a shrub, to ease symptoms or to make other treatments like radiation or chemotherapy more effective, as they have less disease to target post-surgery.

The decision to pursue surgical intervention is multifaceted, involving a judicious look at the tumor's size, type, location, and the patient's health status. For instance, a small, accessible breast cancer tumor might be addressed with a lumpectomy, where only the tumor and a small margin of tissue are removed. Conversely, a large, invasive abdominal tumor may necessitate a more extensive surgical approach, perhaps even a radical resection where entire organs or lymph nodes are removed. The guiding principle in this determination is to achieve maximum benefit with the least possible harm to the patient's quality of life.

Each surgical option comes with its balance of risks and potentials, and understanding them requires a clear view of not just the present state of the disease but the predictive outcomes of the surgery. Advancements in surgical techniques, such as laparoscopic and robotic-assisted surgeries, have fine-tuned the precision and recovery of these procedures, allowing for decisions that are ever more aligned with the patients' well-being and long-term health. This in-depth exploration of surgical strategies underscores their pivotal role in the complex mosaic of cancer treatment.

In the landscape of cancer surgery, the evolution from traditional to minimally invasive techniques marks a significant leap in patient care. Traditional surgeries, typified by substantial incisions, grant surgeons direct and broad access to the tumor. The advantages of such procedures are balanced against longer hospital stays and recovery periods, as well as greater risks of infection and post-surgery pain.

In contrast, minimally invasive surgeries, like laparoscopy, involve small incisions. Surgeons use special instruments and cameras to navigate and remove the tumor with precision. The benefits of these techniques are manifold: shorter hospitalization, less pain, faster return to daily activities, and minimized scarring. Robotic-assisted surgeries go a step further, offering surgeons a high-definition, 3D view of the surgical site and instruments that bend and rotate beyond the human hand's capability, enhancing precision and flexibility.

The choice between these methods hinges on several criteria: the tumor's size and location, the patient's health and preference, and the surgeon's expertise. For instance, a large or unfavorably located tumor may necessitate traditional surgery due to the need for extensive access, whereas smaller, accessible tumors are prime candidates for minimally invasive techniques.

Post-operative considerations profoundly influence surgical decisions. Key amongst these are potential complications and recovery time—factors where minimally invasive and robotic-assisted surgeries often offer superior outcomes. The goal is to ensure not just survival but a better quality of life after treatment.

Furthermore, the role of surgery within a comprehensive treatment plan is nuanced. Neoadjuvant chemotherapy is employed before surgery in some cases to shrink tumors, making them easier to remove or converting an inoperable tumor into a surgically viable one. Post-operative, or adjuvant, radiation therapy may follow surgery to eradicate microscopic remnants of cancer, reducing the risk of recurrence.

Together, these therapies form a cohesive strategy, a coordinated attack against cancer personalized for each patient's specific scenario. Surgical advancements, combined with a deeper understanding of synergy between various treatments, offer a potent blend that is reshaping the horizons of cancer care.

Imagine a sniper, perched with precision, aiming carefully to take out a specific target without causing unnecessary havoc. This is akin to how radiation therapy works in treating cancer. The therapy uses concentrated beams of radiation, like the sniper's focused scope, to hone in on the cancerous cells, damaging their DNA and eliminating their ability to reproduce and spread. Just as a sniper adjusts their aim based on wind and distance, radiation therapy is meticulously planned and delivered using detailed imaging, ensuring the maximum dose reaches the tumor while sparing the surrounding healthy tissue as much as possible.

At its core, radiation therapy disrupts the very blueprint of cancer cells, effectively causing them to self-destruct while leaving neighboring healthy cells intact. It's a delicate balance, like cutting a wire on a bomb to defuse it without setting it off. The technique's precision is crucial; it's the reason why radiation therapy can be such a potent ally in the cancer treatment arsenal and why it's often used in concert with other treatments like surgery and chemotherapy. This interplay of therapies aims to corner cancer on all fronts, offering patients not just a fighting chance, but a strategized road to recovery.

Here is the breakdown of the intricate components involved in radiation therapy, demystifying its application and sophistication in cancer treatment:

- **Types of Radiation Therapy:**
  - **External Beam Radiation:**
    - Intensity-Modulated Radiation Therapy (IMRT): Utilizes multiple beams of different intensities, precisely targeting the tumor while protecting

nearby healthy tissue.

- Proton Beam Therapy: Employs protons instead of X-rays, allowing for a concentrated dose of radiation that stops at the tumor site, thereby reducing damage to surrounding tissues.

- **Brachytherapy:**

- Involves the placement of radioactive material directly inside or adjacent to the tumor, minimizing exposure to healthy tissue while delivering a high radiation dose to the tumor.

- **Intraoperative Radiation:**

- Administered during surgery, it allows direct radiation to the tumor or the tumor bed post-excision, sparing the overlying skin and normal tissues.

- **Treatment Planning Process:**

- **Simulation:**

- A CT scan is conducted to define the precise location and shape of the tumor; this 'simulation' sets the stage for accurate treatment planning.

- **Treatment Planning:**

- Oncologists and dosimetrists collaborate to calculate the exact dose of radiation needed and the angles at which the beams should be aimed to best target the tumor and spare normal tissue.

- **Delivery Methods:**

- **Linear Accelerators:**

- These are high-energy X-ray machines that can conform the radiation beam to the shape of the tumor, delivering radiation from various angles around the patient.

- **Cyberknife:**

- As a form of robotic radiosurgery, this equipment delivers highly precise radiation doses to tumors from virtually any direction with sub-millimeter accuracy.

- **Personalization of Treatment:**

- **Individual Factors:**

- Treatment is customized based on tumor size and location, the particular type of cancer, surrounding sensitive structures at risk, and patient health condition to achieve the best therapeutic ratio.

By understanding the versatility and precision of radiation therapy, one can appreciate its critical place in cancer treatment. Each step, from planning to delivery, is a calculated, evidence-based process aimed at maximizing the eradication of cancer cells while safeguarding the patient's well-being. Through the collaborative efforts of a multidisciplinary team, radiation therapy becomes a fine-tuned instrument in the broader orchestra of cancer care.

Picture chemotherapy as the relentless general in a battle against an army of insurgent cancer cells. This medical strategy dispatches a squad of drugs, akin to soldiers, throughout the body on a search-and-destroy mission, to identify and attack rapidly multiplying cells — which is the hallmark of cancer. Much like a special forces unit, these drugs are trained to pinpoint the enemy, incapacitate them, prevent their replication, and ultimately lead to their elimination.

However, this battalion faces daunting challenges. Since cancer cells often share characteristics with some of the body's healthy cells, chemotherapy can unintentionally harm these innocent bystanders, resulting in the collateral damage we refer to as side effects — from the well-known hair loss to fatigue and vulnerability to infections. Additionally, much like an adaptable enemy, cancer cells can develop resistance to chemotherapy over time, learning to evade or neutralize the attack, rendering the treatment less effective.

Analogous to adapting war tactics, oncologists must modify their strategies — changing drug regimens, dosing, and combination approaches to outmaneuver cancer's defenses. Despite the challenges, when chemotherapy is strategic, it can play a pivotal role in shrinking armies of cancer cells, providing a tactical advantage in the ongoing battle for health. It's a complex, multifront engagement with high stakes, where the goal is not only to win battles but also to increase the chances of winning the war while preserving the well-being of the patient as much as possible.

Here is the detailed breakdown of chemotherapy types, their actions, and strategic roles in treatment:

- **Types of Chemotherapy Drugs:**
  - **Alkylating agents:**
    - Function by binding to DNA, causing crosslinks that prevent cells from replicating, like applying a padlock on the DNA's ability to open and copy itself.
  - **Antimetabolites:**

- Work as imposters, mimicking the normal building blocks of DNA, disrupting the normal process and leading to a breakdown in the replication cycle, akin to supplying the wrong parts to a machine during assembly.
  - **Topoisomerase inhibitors:**
  - Their role is to interfere with topoisomerases, enzymes that help untangle DNA during replication, leading to breaks that impede cell division, much like cutting the power to an essential machine in a production line.

- **Mechanisms of Action:**
  - **Cell cycle specificity:**
  - Some agents specifically attack during certain phases of the cell cycle, targeting the process where cancer cells are most vulnerable, similar to striking an enemy when they're rearming.
  - **DNA interaction:**
  - These drugs chemically modify DNA or interfere with its replication machinery, preventing cancer cells from duplicating. Picture this as a method of jamming the signals that give the order for cell growth.

- **Strategic Use in Treatment Plans:**
  - **Monotherapy:**
  - A solitary drug might be chosen because of its precision and efficacy against certain cancer types, making it the most efficient tool for the job.
  - **Combination therapy:**
  - Combines several drugs, each with a unique method of attack, to outmaneuver cancer's defenses, akin to a coalition force mounting a concerted effort against a common enemy.
  - **Adjuvant and neoadjuvant therapy:**
  - Chemotherapy may precede (neoadjuvant) or follow (adjuvant) other treatments like surgery, to shrink tumors beforehand or clean up any remaining cancer cells afterward, securing the battlefield before and after the main fight.

Each of these components is pieced together with the intent to craft a treatment that maximizes impact on cancer cells while minimizing harm to the patient. Understanding the specifics behind each option provides a clearer picture of how tailored and tactical cancer treatment can be, reflecting the advance in medical strategies that strive for not only survival but also maintaining the quality of life.

Targeted therapy and immunotherapy represent two precision-guided approaches in the vanguard of cancer treatment. Targeted therapy, akin to a locksmith crafting a key specifically designed for a particular lock, involves creating medications that lock onto unique proteins or genes within cancer cells. By identifying and exploiting specific vulnerabilities within the cancer cells, these therapies can disrupt signals that tell the cells to grow or divide, effectively halting the progression of the disease.

Where targeted therapy is the specialist tool, immunotherapy equips the body's natural defenses, training the immune system, much like a coach develops a sports team, to better recognize and combat cancer cells. It can involve a range of treatments, from checkpoint inhibitors that take the 'brakes' off immune cells to allow them to attack cancer, to CAR T-cell therapy, which involves reprogramming a patient's immune cells to seek and destroy cancer cells.

While these treatments are powerful, they come with a caveat. Targeted therapies often require the presence of very specific molecules within the cancer cells, making their scope powerful yet limited. Immunotherapies, meanwhile, can sometimes overly activate the immune system leading to significant inflammation or tissue damage, akin to a bodyguard overzealously protecting someone.

These advanced cancer treatments are reshaping the landscape of patient care, allowing for more personal and effective approaches. By providing a granular understanding of how these therapies work individually, the reader can appreciate the elegance of these medical innovations and their transformative impact on cancer care.

Let's dive into the intricacies of targeted therapies and immunotherapies, unveiling their precise roles in contemporary cancer treatment:

- **Subtypes of Targeted Therapies:**
  - **Common Drugs:**
    - HER2 inhibitors, such as trastuzumab, work in breast cancer patients by attaching to the HER2 protein on cancer cells, which blocks the cells from receiving growth signals.
    - Tyrosine kinase inhibitors, like imatinib, target specific enzymes (tyrosine kinases) in chronic myeloid leukemia cells that trigger growth signals, effectively switching off their proliferative capacity.

- **Biomarker Testing:**
- Tests identify molecular 'flags', known as biomarkers, unique to a person's cancer, determining the suitability of specific therapies, much like a scanner that reads a barcode to match a product.

- **Categories of Immunotherapies:**
- **Active vs. Passive:**
- Active immunotherapies, like cancer vaccines, work by stimulating the immune system to attack cancer cells more effectively.
- Passive immunotherapies involve monoclonal antibodies that are created outside the body and administered to patients to help the immune system recognize and fight cancer cells.
- **Examples:**
- Immune checkpoint inhibitors, such as pembrolizumab, release the 'brakes' on immune cells, allowing them to attack melanoma cells more vigorously.

- **Treatment Protocols and Management:**
- **Administration Routes:**
- Treatments may be given intravenously, introducing them directly into the bloodstream, or orally in pill form that circulates through the body to find and impact cancer cells.
- **Course Durations:**
- Treatment schedules can vary, with some drugs given daily and others in cycles, with a period of treatment followed by a rest period to allow the body time to recover.
- **Follow-Up Care:**
- After therapy, regular monitoring, through methods such as MRI or CT scans and blood tests like tumor markers, assesses the effectiveness of therapy and dictates further treatment needs.

Getting to know the particularities of these therapies offers a panoramic view of their application in medicine. Battling cancer is as much about the strategic use of these remarkable treatments as it is about their molecular mechanics. By understanding the fine points of how they operate and are managed, one can comprehend their profound effect on both the disease and the human story behind each patient.

Personalized medicine in oncology is an approach that tailors patient treatment based on the individual's genetic makeup, particularly that of their tumor cells. The process begins with genetic profiling, which can be likened to creating a detailed map of a tumor's DNA. Just as a navigator uses maps to chart a course, oncologists use the information from genetic profiling to identify aberrations or mutations that are the hallmarks of a patient's cancer.

Armed with this information, healthcare professionals can formulate treatment strategies that target those specific genetic changes. For instance, if genetic profiling reveals a mutation in a protein that drives tumor growth, a therapy that specifically inhibits that protein can be selected. This approach stands in stark contrast to traditional cancer treatments, which are applied broadly to all patients with a certain type or stage of cancer.

However, while personalized medicine holds immense promise, it also faces limitations. Not all patients will have a recognizable genetic target for which there is an available therapy, and the nature of cancer to evolve and adapt can render targeted treatments less effective over time. Furthermore, genetic profiling requires highly specialized laboratory capabilities that may not be accessible in all healthcare settings.

Despite these challenges, the advances in personalized oncology ignite hope. This strategy paves the way for more effective treatments with potentially fewer side effects since it concentrates on the patient's distinct genetic profile. By leveraging the power of genetic profiling, oncology is shifting towards more precise, informed, and individualized patient care.

The molecular footprint of a patient's cancer is the starting point for the personalization of their treatment. Genetic profiling unfolds in several meticulous steps, beginning with a biopsy, where a sample of the cancer tissue is surgically obtained. The DNA is then extracted from the cells within this tissue, providing the raw material for genetic analysis.

The sequencing of this DNA traditionally employs next-generation sequencing (NGS) technologies, which can rapidly read and compile the vast sequences of genetic code. This process effectively sifts through the genetic information to spot mutations—irregularities in the DNA that could drive or influence the growth of cancer cells. These mutations are then mapped to understand their functional implications, creating an accurate genetic landscape of the individual's cancer.

With this map, oncologists can navigate the complex terrain of treatment options. Molecular targets identified during profiling inform the selection of targeted therapies—drugs designed to seek out and disable the genetic errors promoting cancer growth. Similarly, recognized mutations can indicate which immunotherapy might be most effective, guiding the immune system to recognize and destroy cancer cells bearing these specific molecular flags.

Indeed, algorithms aid this decision-making by weighing a combination of genetic markers, the evidence for each potential therapy, and individual patient factors to recommend the most appropriate treatment. Such algorithms assimilate vast data into a coherent strategy, illuminating the path forward in what would otherwise seem a genetic labyrinth.

Monitoring the treatment's effectiveness is an ongoing process, akin to vigilant surveillance. By continually assessing how the cancer responds, often with follow-up genetic profiling or through imaging techniques, doctors can adapt the strategy, adding, changing or halting treatments in response to the cancer's evolution.

Personalized medicine stands as a pillar in modern oncology, a beacon of progress that channels the power of genetics into targeted, effective care. As this process becomes even more refined, the hope rises for treatments that are not only more successful but also more tailored to the individual, preserving the essence of who they are while combating the disease that threatens them.

Alex Trebek and Ruth Bader Ginsburg, though from different walks of public life, shared the personal and arduous challenge of battling cancer under the public lens. Trebek, the beloved "Jeopardy!" host, faced pancreatic cancer with the same grace and candor he brought to TV screens for decades. His openness about treatments, which included surgery and rounds of chemotherapy, allowed a rare insight into the resilience required in the face of such an unpredictable disease. Trebek's brave fight undeniably brought pancreatic cancer into the discourse, shedding light on the importance of early detection and research.

Justice Ruth Bader Ginsburg, on the other end of the spectrum, battled cancer multiple times. From colorectal to pancreatic cancer, she underwent surgeries, chemotherapy, and radiation. Her multiple public victories over cancer not only highlighted the advancements in treatment but also symbolized her relentless spirit. Both Ginsburg and Trebek showcased incredible fortitude and, in doing so, generated public dialogue about the critical need for continued support in cancer research and the value of tireless advocacy.

Their journeys, while immensely personal, became a platform for advocacy and awareness. With each update, they inadvertently educated the public about the complexities of treatment, the reality of side effects, and the often-overlooked emotional journey of a cancer patient. By simply sharing their own stories, they have inspired countless others to engage with their health proactively and supported a wider culture of openness around the cancer experience. Their legacy in the realm of cancer awareness is as enduring as their professional contributions, serving as a beacon of hope and igniting a collective push for advancements in care and cure.

The landscape of cancer treatment has undergone transformative change, evolving significantly from the one-size-fits-all approach of the past to today's era of personalized medicine. Treatment strategies are no longer dictated solely by cancer type and stage but are increasingly influenced by the intricate genetic makeup of both the patient and the tumor. Genetic profiling, which maps the unique genetic aberrations in cancer cells, now guides the selection of therapies, allowing for an unprecedented level of customization in patient care.

This evolution brings with it complexities—the necessity of advanced laboratory resources, the challenge of keeping up with rapidly expanding knowledge, and the enduring quest to address the heterogeneity of cancer response to treatment. Yet, it is these very complexities that fuel progress, driving the continuous search for more sophisticated and effective therapies.

The advances in cancer treatment extend the beacon of hope, shining brighter with each breakthrough that allows patients to receive more targeted, less toxic, and potentially more effective therapies. As scientists and clinicians push the boundaries of what is possible in oncology, patients worldwide are empowered with an expanding arsenal in the battle against cancer and a future that holds the promise of lifetimes extended and quality of life enhanced. This summary not only acknowledges the milestones achieved but also honors the collective endeavor to ensure that these advancements reach all who stand to benefit, transforming hope into a tangible reality for patients across the globe.

CHAPTER 8: LIVING WITH CANCER

Facing cancer is like riding the most daunting roller coaster imaginable. It begins with the steep climb of anxiety and fear, akin to the feeling one experiences while ascending the coaster's first peak, the world below growing distant and distorted. The moment of diagnosis marks the precipice—the brief pause before life plunges into the unknown at breakneck speed.

Treatment brings a series of lows and highs, with moments of hope that surge like the coaster cresting a rise, offering a breath of relief and a burst of elation. Yet, these are interspersed with the dread of descent, where the gravity of the situation pulls one down into the valleys of pain, fatigue, and uncertainty—the tightening of the stomach as the coaster hurtles downward.

Remission, when achieved, is the reprieve of a level track, a straight path offering respite and a chance to catch one's breath. But the journey may twist again, compelling the rider to brace against the gnawing fear of recurrence, akin to the coaster's sudden, jolting turns. Through it all, emotions ebb and flow, wax and wane, with each twist and turn of the experience demanding a resilience that is as endearing as it is challenging.

This emotive voyage of the cancer journey captures not just the physical tribulations but the mental fortitude required to endure and overcome. It speaks to the heart of what it means to confront one's mortality and to wrestle with the profound truths that emerge in life's most unexpected moments, all while maintaining the courage to continue the ride.

Here is the breakdown on the emotional stages following a cancer diagnosis and during treatment, complete with coping strategies and the importance of support systems:

- **Initial Diagnosis:**
  - **Shock and Denial:**
    - Utilizing temporary disengagement for initial coping

- Engaging in mindful breathing exercises to manage immediate stress responses
  - **Acceptance:**
    - Journaling or expressive writing to acknowledge and process feelings
    - Consulting with therapists or counselors experienced in oncology to facilitate acceptance

- **Treatment Phase:**
  - **Anxiety and Hope:**
    - Adopting relaxation techniques such as meditation and guided imagery
    - Setting small, achievable goals to instill a sense of progress and hope
  - **Coping with Side Effects:**
    - Establishing a routine for rest and self-care to combat physical fatigue
    - Participating in support groups for shared experiences and tips

- **Remission and Beyond:**
  - **Embracing Normalcy:**
    - Gradually resuming work and social activities to reestablish normal life patterns
    - Exploring new hobbies or interests to create a fresh sense of purpose and engagement
  - **Vigilance Against Recurrence:**
    - Staying informed about signs of recurrence and maintaining regular follow-up appointments
    - Practicing mindfulness and cognitive restructuring to manage fears constructively

- **Role of Support Systems:**
  - **Family and Friends:**
    - Providing emotional support through companionship and active listening
    - Offering practical assistance with daily tasks and treatment logistics
  - **Professional Support:**
    - Accessing mental health professionals for strategies to cope with the psychological toll of cancer
    - Joining structured support groups to connect with others and share coping mechanisms

This comprehensive layout presents each emotional phase of the cancer journey with empathetic understanding, equipping readers with the resources and knowledge to navigate this challenging period. The discussion translates complex emotional responses into actionable strategies, emphasizing their role in cultivating resilience and maintaining quality of life. These insights offer a foundation of support, reinforcing the idea that while the path may be difficult, no one has to walk it alone.

A cancer diagnosis acts as a seismic shift in one's social landscape, redefining interactions and the composition of support networks. Following such a diagnosis, an individual may notice changes in their relations; some friends may unexpectedly step back, uncomfortable with the illness, while others move closer, offering a shoulder and practical aid. Family dynamics can alter too, with roles shifting to accommodate new caregiving responsibilities in a manner reminiscent of a community rallying after a natural disaster.

Support systems amplify in importance, becoming akin to a vital utility network that keeps the home fires burning. New connections often form, such as support groups—face-to-face or online—serving as forums for collective experience sharing and encouragement, mirroring group structures often found in recovery programs. Professional help may come into play, with therapists and counselors specializing in oncology providing guidance and strategies for coping with the psychological aftermath of the disease, similar to disaster relief services following a calamity.

Recognizing the presence and value of these social modifications is critical. Just as businesses re-strategize in changing markets, individuals and families must adapt to their new societal framework post-diagnosis, identifying which support channels to bolster and where to seek new connections. This agile reconfiguration of social and support pillars is pivotal not just for resilience but for maintaining one's identity and agency in the face of cancer's challenges. This exploration elucidates the underpinnings of how relationships evolve on this journey, highlighting the integral role community plays in the healing process.

Let's take a closer look at the nuanced shifts that can occur in the realm of personal relationships and community networks following a cancer diagnosis:

- Relationship Adjustments:

- Shifts in friendships often materialize, where some friends may reel from the news and retreat, leading to less frequent contact. Alternatively, some people unexpectedly become stalwarts, readily engaging in activities like meal preparation or accompanying the individual to treatments, providing both tangible help and emotional sustenance.

- Family dynamics are likely to evolve as roles adjust in response; a spouse may assume more home management duties, or children might take on additional responsibilities, like caring for younger siblings. The family unit often redefines itself around the needs of the affected member, sometimes leading to stronger familial bonds through shared adversity.

- Support Groups:

- The landscape of support groups is diverse, encompassing structured, professionally-led groups focused on specific types of cancer, which might provide medical information and coping strategies, to more informal peer-led assemblies in community centers or places of worship, offering a space for sharing personal experiences and mutual encouragement.

- These groups become pillars of emotional support, with discussions that often cover the spectrum from navigating healthcare systems to managing daily life challenges. People find solace in the shared narratives and the collective wisdom that emerges in these supportive settings.

- Professional Counselors:

- Oncology mental health professionals occupy a critical role, offering counseling tailored to the challenges of living with cancer. They employ approaches such as cognitive-behavioral therapy to help manage anxiety and mood disorders, or they may provide existential therapy to aid individuals in finding meaning following a cancer diagnosis.

- Goals of such counseling typically include the enhancement of coping skills, the mitigation of distress, and the facilitation of communication with loved ones, all aiming to improve the patient's overall quality of life.

- Community Integration:

- Communities often rally in support, akin to a reflexive embrace of one of their own. Neighbors might organize meal rotations, provide transportation for family members, or help with household chores, exemplifying grassroot community support.

- Larger community initiatives can also be seen in action, such as fund-raising events or awareness campaigns, which serve dual roles in providing material support and emboldening the patient's spirit through a demonstration of collective care and concern.

These detailed components lend clarity to the social and systemic changes that manifest when cancer enters an individual's life. Recognizing each strand in this social fabric illuminates the importance of human connection and the myriad ways it intertwines to form a strong and nurturing network, integral to navigating the cancer experience.

Imagine a building, standing resolute against the elements. Its strength lies not just in the visible bricks and mortar, but in its unseen skeleton – the beams and columns that prop it up from within. Support networks for those with cancer are much like these foundational structures, often invisible but utterly vital to the integrity of the whole.

Family and friends form the pillars, bearing the weight of practical needs and emotional strains with steadiness and love. Just as load-bearing walls distribute the stress to prevent collapse, these human pillars lend their shoulders to divide the burden a cancer diagnosis carries. They provide the crucial stability that allows daily life to carry on in the face of adversity.

Professional support groups and caregivers are akin to the framework that holds the building's shape. They bring expertise and shape the healing process. Like the scaffolding that supports a structure during its construction and maintenance, they offer the necessary tools and resources to navigate treatment and recovery, bolstering the patient through the stages of their journey.

The community at large acts as the architectural envelope – the settings and systems that encase and protect. From community-led fundraisers to advocacy networks and beyond, this broader support mirrors the collective work of a society constructing safety nets and fostering environments conducive to healing and hope.

Just as a building relies on various kinds of support to remain upright, so too do individuals with cancer rely on diverse networks to face their challenges. These connections are as crucial to the emotional and psychological durability of a person as the foundations are to the physical edifice. This view not only clarifies the nature of these support systems but also celebrates their role in weaving a web of care that can catch and cradle those who fall into its folds.

Here is the breakdown on the specific roles within support networks for cancer patients, showcasing their invaluable contributions:

- **Family as Pillars:**
  - **Emotional Pillars:**
    - Listen empathetically to concerns and fears without judgment, acting as the emotional bedrock in uncertain times.
    - Offer reassurance and encouragement, fostering a positive environment that promotes healing.
  - **Practical Pillars:**
    - Manage logistics like scheduling and transportation to medical appointments, ensuring consistency in treatment.
    - Assist with medication regimens and daily task management, helping maintain a semblance of normalcy.
    - Help navigate the complexities of insurance and financial management, mitigating additional stressors for the patient.

- **Professional Support Groups and Caregivers:**
  - **Expert Framework:**
    - Deliver informed advice on treatment options and side effects, acting as a guiding light through the medical maze.
    - Provide psychological support tailored to cancer patients' unique needs, helping anchor them emotionally.
  - **Scaffolding during Recovery:**
    - Conduct regular counseling sessions, forming the backbone of emotional recovery.
    - Offer personalized care routines, such as physiotherapy or nutritional guidance, to rebuild the patient's strength.

- **Community Integration:**
  - **Protection and Encapsulation:**
    - Organize fundraisers and health fairs, contributing to an envelope of community care.
    - Advocate for patient rights and better healthcare policies, striving to secure a more resilient healthcare framework.
  - **Constructive Efforts:**

- Establish services like meal delivery or child care assist, directly nurturing the patient's and family's wellbeing.
- Coordinate visitation schedules or 'buddy' systems, weaving a fabric of presence and support for those who might feel isolated.

This narrative underscores the multidimensional support structures that embrace cancer patients. Each role, performed with dedication and empathy, cements the broader foundation of care – serving as a testament to how this support network is as pivotal to the emotional and functional stability of a patient as the foundational supports to a building's integrity.

Maintaining quality of life for a cancer patient involves a multi-pronged strategy that targets not only the physical symptoms but also the psychological impact of the illness. Pain management, one of the cornerstones of this strategy, often employs medications, but it can also include other therapies like massage, acupuncture, or cold and heat application—accessible techniques reminiscent of common remedies for everyday aches.

To combat fatigue, which is common during cancer treatment, patients are encouraged to balance rest with activities that energize them, such as short walks or light exercise—akin to recharging one's batteries to keep the machinery running smoothly. Nutritional support is another critical component, with diet modifications tailored to help mitigate treatment side effects and maintain strength, much like using the right fuel to optimize the performance of a vehicle.

Emotional well-being is addressed through regular consultations with mental health professionals. Strategies such as talk therapy or cognitive-behavioral therapy can unpack and manage the profound emotions that accompany cancer, in a setting as candid and constructive as a heart-to-heart with a trusted confidant. Peer support groups also provide a shared space for connection and understanding, offering the sense of camaraderie found in team sports where individuals unite under a common goal.

In this straightforward explanation, each strategy is an essential gear in the broader mechanism of care designed to sustain cancer patients, both physically and emotionally. By ensuring these gears operate in concert, the aim is to enable patients to remain in the driver's seat of their lives, making informed decisions that reflect their values and preferences, even amidst the complexities of cancer.

Managing the quality of life for cancer patients is crucial, and requires a personalized approach. Here are the steps that ensure tailored support across several key areas:

1. **Personalized Pain Management Plan:**
- Begin with a thorough assessment of the patient's pain levels and sources.
- Develop a pain control plan incorporating medications, tailored to the patient's specific type of pain and medical history.
- Integrate alternative therapies such as acupuncture, massage, or mindfulness meditation to complement the medical approach.
- Regularly review and adjust the plan based on the patient's feedback and any changes in their condition.

2. **Assessing and Addressing Fatigue:**
- Monitor the patient's energy levels and patterns of fatigue through daily tracking.
- Collaborate with the patient to identify periods of rest and light activities that align with their energy peaks and troughs.
- Gradually introduce or modify exercise routines to match the patient's evolving capacity, with a focus on low-impact exercises like walking or gentle yoga.

3. **Creating a Dietary Regimen:**
- Conduct a comprehensive nutritional assessment, taking into account the patient's treatment side effects, tastes, and any dietary restrictions.
- Design a meal plan rich in the necessary vitamins and proteins to rebuild strength, while being mindful of foods that can help alleviate side effects such as nausea.
- Schedule regular consultations with a dietitian to refine the meal plan and ensure it continues to meet the patient's nutritional needs.

4. **Ongoing Mental Health Support:**
- Spearhead a discussion about the patient's emotional well-being to identify specific areas of concern.
- Provide information on accessing professional therapy services,

including individual, group, or family sessions.

- Encourage enrollment in peer support groups for shared understanding and emotional connection.

- Set up a routine for mental health checkpoints to ensure the patient feels continuously supported.

5. **Synchronizing Various Support Strategies:**

- Establish a central coordination point, possibly a case manager or primary caregiver, to ensure all aspects of care are integrated and communicated among the patient's healthcare team.

- Schedule regular multidisciplinary team meetings to discuss the patient's progress across all areas of the support plan.

- Adjust different strategies in tandem, ensuring changes in one area, such as diet, are reflected and supported in other areas, like energy management.

Following these detailed steps means embracing a holistic view of cancer care. It's not simply about treating the disease; it's about nurturing the person living with it—mind, body, and spirit—as a unified whole. This approach ensures each part of the care plan works together, like a symphony, each note contributing to the harmony of the patient's life.

The journeys of Lance Armstrong and Stephen Jay Gould through their cancer battles serve as beacons in the night for those navigating the murky waters of their own cancer experiences. Armstrong's triumph over testicular cancer to claim seven Tour de France titles is a vivid image of human endurance, painting the image of a downed cyclist who not only gets back on his bike but races to victory against all odds. It is a living manifesto that physical ailments do not define limits but can fuel an indomitable will to overcome.

Stephen Jay Gould's narrative, infused with the intellect of a seasoned scientist, offers a different, yet equally powerful type of inspiration. In his essay "The Median Isn't the Message," Gould takes the cold statistics of cancer prognosis and imbues them with a personal significance. His story is a lighthouse for rational hope amidst the fog of statistical averages, providing an analytical framework to view one's prognosis not as a verdict but as one variable in the complex equation of life.

These storied paths resonate with familiarity, reminding us of universal tales of the underdog and the sage, providing more than just instruction; they offer kinship. In sharing their sagas, they knit the cancer community closer, embedding the lesson that while cancer is a formidable adversary, the narratives we author about our battles can sustain morale, cultivate grit, and fortify the collective spirit against despair.

Here is the breakdown of key elements that underline the inspirational power of personal narratives in the face of cancer, as exemplified by Lance Armstrong and Stephen Jay Gould:

- Inspirational Elements of Lance Armstrong's Narrative:
- The sheer force of will exhibited in overcoming a grave prognosis to not only survive but to excel professionally in cycling, defying medical odds.
- Armstrong's leveraging of his celebrity to foster broader change, exemplified by founding a support organization that provides resources and advocacy for cancer communities.
- His athletic accomplishments resonate as a representation of conquest over adversity, providing an embodied example of victory over life's most challenging obstacles.

- Inspirational Elements of Stephen Jay Gould's Narrative:
- Gould's interpretation of cancer survival statistics as a discourse of hope, transforming cold data into a beacon of possibility for individual patients.
- The infusion of his expertise as a paleontologist into his cancer narrative, which lends a unique perspective on time, survival, and the narrative of illness.
- Gould's narrative serves as intellectual solace, offering those grappling with diagnosis an analytical framework to contextualize their situation and reclaim a sense of agency.

- General Impact of Personal Narratives:
- Storytelling serves as a psychological bolster, enhancing emotional resilience by providing examples of strength and recovery that individuals can relate to and draw strength from.
- Relatable stories enhance visibility, creating a image of shared experiences that contribute to a supportive and understanding community for those undergoing similar battles.

This analysis reveals personal narratives as more than tales of individual struggle; they are the threads that weave a image of collective resilience and unity, reminding those in the midst of their fight against cancer that they are not alone and that their stories, too, can inspire and console others.

Chapter 8: Living with Cancer brings to light the multifaceted existence that comes with a cancer diagnosis. It traverses the challenging landscape marked by fear, adjustments in daily life, and the discovery of inner strength. This chapter delves into the emotional whirlwind, from the gut-punch of the initial diagnosis to the resolve that forms from ongoing battles and small victories. It lays bare the changes in personal dynamics, noting how relationships may evolve, with some providing unexpected bastions of support, taking on roles similar to scaffolding in times of reconstruction.

As much as cancer is a personal battle, it is also a communal concern. The chapter recognizes how support groups, friends, and even strangers join forces to create a collective embrace, often instrumental in uplifting spirits. It underscores the integral role of advocacy, instilling the power of informed voices in demanding better care, advanced research, and policies that protect patient interests.

Each theme within this chapter is an affirmation of life's continuity amidst adversity and a call to arms for understanding and advocacy. It imparts that living with cancer is more than enduring treatment; it's about maintaining one's essence, nurturing hope, and drawing from the deep wells of community support. This is no solitary journey but a shared experience, where knowledge, compassion, and action converge to make a tangible difference.

CONCLUSION

As we close the final pages of "Cancer Explained," we reflect on a journey through the labyrinthine world of a disease that affects millions yet remains shrouded in mystery for far too many. From the microscopic mutation of cells to the global charge for a cure, this book has aimed to equip its readers with a comprehensive understanding of cancer, translating medical complexities into clear, manageable knowledge.

Key themes have emerged—of resilience in the face of uncertainty, of innovation at the frontier of science, and of hope that thrives amidst adversity. You've seen how cancer's reach extends beyond the individual, influencing family dynamics, community structures, and healthcare systems worldwide. The book has underscored the importance of early detection, the promise held by new treatments, and the power of personalized medicine tailored to each unique battle.

"Cancer Explained" has woven together not just facts and figures but stories of defiance and triumph that embody the human spirit. The knowledge imparted here serves as a shield and a beacon: a shield against fear borne from the unknown, and a beacon guiding towards informed choices, advocacy, and the continual pursuit of a life unfettered by cancer's grasp.

As you step away from this text, carry forward the lessons learned and the perspectives gained. Let them inform your conversations, guide your decisions, or perhaps even inspire your contributions to the ongoing fight against cancer. The battle is as complex as the disease itself, but armed with understanding, each of us holds the power to make a difference—in our own lives and in the lives of others. From prevention to treatment, from the support we offer to the policies we champion, every action counts.

Remember, knowledge is the ally of health, and compassion its faithful companion. With these in hand and heart, the journey continues towards a future brighter than the shadow cast by cancer.

ABOUT THE AUTHOR

Jon Adams brings a wealth of experience from over twenty years in the information technology industry, having worked with some of the world's leading tech giants.

With a deep-seated passion for science, technology, and languages, Jon excels at demystifying complex subjects, making them accessible and engaging to a broad audience.

His writings focus on breaking down intricate topics into everyday terms, helping readers not just learn but also apply this knowledge in their daily lives.

Currently, Jon is a proud member of Green Mountain Publishing, which publishes his insightful books. Through his work, he aims to foster a deeper understanding and appreciation of technology and science, enriching readers' lives.

Jon@GreenMountainComputing.com

Made in United States
Orlando, FL
20 July 2025

63138756R00056